Fermat's Last Theorem

CREDIT: CORBIS-BETTMANN

Pierre de Fermat (1601–1665).

FERMAT'S LAST THEOREM

Unlocking the Secret of an Ancient Mathematical Problem

By Amir D. Aczel

Four Walls Eight Windows
New York/London

Published in the United States by:
Four Walls Eight Windows
39 West 14th Street, room 503
New York, N.Y., 10011

First printing October 1996.

Library of Congress Cataloging-in-Publication Data:
Fermat's Last Theorem: unlocking the secret of an ancient mathematical problem / by Amir D. Aczel.
p. cm.
ISBN: 1-56858-077-0
1. Fermat's last theorem. I. Title.
QA244.A29 1996
512'.74—dc20 96-9029
CIP

10 9 8 7 6 5 4 3 2 1

Printed in the United States

To my father

Preface

In June 1993, my old friend Tom Schulte was visiting me in Boston from California. We were sitting at a sunny sidewalk café on Newbury Street, tall, icy drinks in front of us. Tom had just gotten divorced, and he was ruminative. He half turned toward me. "By the way," he said, "Fermat's Last Theorem has just been proved." This must be a new joke, I thought, as Tom's attention was back on the sidewalk. Twenty years earlier, Tom and I were roommates, both of us undergraduate students in mathematics at the University of California at Berkeley. Fermat's Last Theorem was something we often talked about. We also discussed functions, and sets, and number fields, and topology. None of the math students slept much at night, since our assignments were so difficult. That's what distinguished us from students in most other areas. Sometimes we'd have math nightmares . . . trying to prove some theorem or another before it was due in the morning. But Fermat's Last Theorem? No one ever believed it would be proven in our lifetime. The theorem was so difficult, and so many people had tried to prove it for over three hundred years. We were well aware that entire branches of mathematics had been developed as the result of attempts to prove the theorem. But the attempts failed, one by one. Fermat's Last Theorem had come to symbolize the unattainable. I once even used the theorem's perceived impossibility to my advantage. It was a few years later, also at Berkeley, when I had already graduated in math and was getting my Master's in operations research. An arrogant graduate student in mathematics, unaware of my own background in math, offered me his help when we met at the International House where we both lived. "I'm in pure mathematics," he said. "If you ever have a math problem you can't solve, feel free to ask me." He was about to leave, when I said "Um, yes. There is something you can help me with . . . " He turned back, "Why, sure, let me see what it is." I pulled open a napkin—we were at the dining room. I slowly wrote on it:

> $x^n + y^n = z^n$ has no whole number solution when n is greater than 2.

"I've been trying to prove this one since last night," I said, handing him the napkin. I could see the blood draining from his face. "Fermat's Last Theorem," he grunted. "Yes," I said, "You're in pure math. Could you help me?" I never saw that person from up close again.

"I am serious," Tom said, finishing his drink. "Andrew Wiles. He proved Fermat's Last Theorem in Cambridge last month. Remember that name. You'll hear it a lot." That night, Tom was in the air, flying back to California. In the next months, I realized Tom didn't play a joke on me, and I followed the sequence of events where Wiles was at first applauded, then the hole in his proof was found, then he withdrew for a year, and finally reemerged with a corrected proof. But following the continuing saga, I learned that Tom was wrong. It wasn't Andrew Wiles's name I should have paid attention to, or at least not his alone. I, and the world, should have recognized that the proof of Fermat's Last Theorem was far from the work of one mathematician. While Wiles got much of the praise, the accolades belong to others as much: Ken Ribet, Barry Mazur, Goro Shimura, Yutaka Taniyama, Gerhard Frey, and others. This book tells the entire story, including what happened behind the scenes and out of the view of the media's cameras and floodlights. It is also a story of deception, intrigue, and betrayal.

"Perhaps I could best describe my experience of doing mathematics in terms of entering a dark mansion. You go into the first room and it's dark, completely dark. You stumble around, bumping into the furniture. Gradually, you learn where each piece of furniture is. And finally, after six months or so, you find the light switch and turn it on. Suddenly, it's all illuminated and you can see exactly where you were. Then you enter the next dark room . . . "

This is how Professor Andrew Wiles described his seven-year quest for the mathematicians' Holy Grail.

Just before dawn on June 23, 1993, Professor John Conway approached the darkened mathematics building on the Princeton University campus. He unlocked the front door and quickly walked up to his office. For weeks preceding his colleague Andrew Wiles' departure for England, persistent but unspecific rumors had been circulating in the world's mathematical community. Conway was expecting something important to happen. Exactly what it was, he had no idea. He turned on his computer and sat down to stare at the screen. At 5:53 AM, a terse e-mail message flashed from across the Atlantic: "Wiles proves F.L.T."

Cambridge, England, June 1993

Late in June of 1993, Professor Andrew Wiles flew to England. He was returning to Cambridge University, where he had been a graduate student twenty years earlier. Wiles' former doctoral thesis adviser at Cambridge, Professor John Coates, was organizing a conference on Iwasawa Theory—the particular area within number theory in which Andrew Wiles did his dissertation and about which he knew a great deal. Coates had asked his former student if he would mind giving a short, one-hour talk at the conference on a topic of his choice. To his great surprise and that of the other conference organizers, the shy Wiles—previously reluctant to speak in public—responded by asking if he could be given three hours of presentation.

The 40-year-old Wiles looked the typical mathematician when he arrived in Cambridge: white dress shirt with sleeves rolled up carelessly, thick horn-rimmed glasses, unruly strands of thinning light hair. Born in Cambridge, his return was a very special kind of homecoming—it was the realization of a childhood dream. In pursuit of this dream, Andrew Wiles had spent the last seven years of his life a virtual prisoner in his own attic. But he hoped that soon the sacrifice, the years of struggle and the long hours of solitude would end. Soon he might be able to spend more time with his wife and daughters, of whom he had seen so little for seven years. He had often failed to show up for lunch with his family, missed afternoon tea, barely made it to dinner. But now the accolades would be his alone.

The Sir Isaac Newton Institute for Mathematical Sciences at Cambridge had only recently opened by the time Professor Wiles arrived to deliver his three hour-long lectures. The Institute is spacious, set in scenic surroundings at some distance from the University of Cambridge. Wide areas outside the lecture halls are furnished with plush, comfortable chairs, designed to help facilitate the informal exchange of ideas among scholars and scientists, and to promote learning and knowledge.

Although he knew most of the other mathematicians who came to the specialized conference from around the world, Wiles kept to himself. When colleagues became curious about the length of his scheduled presentation, Wiles would only say they should come to his lectures and find out for themselves. Such secretiveness was unusual, even for a mathematician. While they often work alone trying to prove theorems and are generally not known to be the world's most gregarious people, mathematicians usually share research results with each other. Mathematical results are freely circulated by their authors in the form of research preprints. These preprints bring their authors outside comments that help them improve the papers before they are published. But Wiles didn't hand out preprints and didn't discuss his work. The title of Wiles' talks was "Modular Forms, Elliptic Curves, and Galois Representations," but the name gave no hint where the lectures would lead, and even experts in his field could not guess. The rumors intensified as time went on.

On the first day, Wiles rewarded the 20 or so mathematicians who came to his lecture with a powerful and unexpected mathematical result—and there were still two more lectures to go. What was coming? It became clear to everyone that Wiles' lectures were the place to be, and the suspense grew as expectant mathematicians flocked to the lectures.

On the second day, Wiles' presentation intensified. He had brought with him over 200 pages of formulas and derivations, original thoughts stated as new theorems with their lengthy, abstract proofs. The room was now filled to capacity. Everyone listened intently. Where would it lead? Wiles gave no hint. He dispassionately continued writing on the blackboard and when he was done for the day, he quickly disappeared.

The next day, Wednesday, June 23, 1993, was his last talk. As he neared the lecture hall, Wiles found it necessary to push his way in. People stood outside blocking the entrance and the room was overflowing. Many carried cameras. As Wiles again wrote seemingly endless formulas and theorems on the board, the tension increased. "There was only one possible climax, only one possible end to Wiles' presentation," Professor Ken Ribet of the University of California at Berkeley later told me. Wiles was finishing the last few lines of his proof of an enigmatic and complicated conjecture in mathematics, the

Shimura-Taniyama Conjecture. Then suddenly he added one final line, a restatement of a centuries-old equation, one which Ken Ribet had proved seven years earlier would be a consequence of the conjecture. "And this proves Fermat's Last Theorem," he said, almost offhandedly. "I think I'll stop here."

There was a moment of stunned silence in the room. Then the audience erupted in spontaneous applause. Cameras flashed as everyone stood up to congratulate a beaming Wiles. Within minutes, electronic mail flashed and faxes rolled out of machines around the world. The most celebrated mathematical problem of all time appeared to have been solved.

"What was so unexpected was that the next day we were deluged by the world press," recalled Professor John Coates, who organized the conference without having the slightest idea that it would become the launching ground for one of the greatest mathematical achievements. Headlines in the world's newspapers hailed the unexpected breakthrough. "At Last, Shout of 'Eureka!' In Age-Old Math Mystery" announced the front page of the *New York Times* on June 24, 1993. The *Washington Post* called Wiles in a major article "The Math Dragon-Slayer," and news stories everywhere described the person who apparently solved the most persistent problem in all of mathematics, one that had defied resolution for over 350 years. Overnight, the quiet and very private Andrew Wiles became a household name.

Pierre de Fermat

Pierre de Fermat was a seventeenth-century French jurist who was also an amateur mathematician. But while he was technically an "amateur" since he had a day job as a jurist, the leading historian of mathematics E. T. Bell, writing in the early part of the twentieth century, aptly called Fermat the "Prince of Amateurs." Bell believed Fermat to have achieved more important mathematical results than most "professional" mathematicians of his day. Bell argued that Fermat was the most prolific mathematician of the seventeenth century, a century that witnessed the work of some of the greatest mathematical brains of all time.[1]

One of Fermat's most stunning achievements was to develop the main ideas of calculus, which he did thirteen years before the birth of Sir Isaac Newton. Newton and his contemporary Gottfried Wilhelm von Leibniz are jointly credited in the popular tradition with having conceived the mathematical theory of motion, acceleration, forces, orbits, and other applied mathematical concepts of continuous change we call calculus.

Fermat was fascinated with the mathematical works of ancient Greece. Possibly he was led to his conception of calculus ideas by the work of the classical Greek mathematicians Archimedes and Eudoxus, who lived in the third and fourth centuries B.C., respectively. Fermat studied the works of the ancients—which were translated into Latin in his day—in every spare moment. He had a full-time job as an important jurist, but his hobby—his passion—was to try to generalize the work of the ancients and to find new beauty in their long-buried discoveries. "I have found a great number of exceedingly beautiful theorems," he once said. These theorems he would jot down in the margins of the translated copies of ancient books he possessed.

Fermat was the son of a leather merchant, Dominique Fer-

Arithmeticorum Lib. II. 85

teruallo quadratorum, & Canones iidem hic etiam locum habebunt, vt manifestum est.

QVÆSTIO VIII.

PROPOSITVM quadratum diuidere in duos quadratos. Imperatum sit vt 16. diuidatur in duos quadratos. Ponatur primus 1 Q. Oportet igitur 16 − 1 Q. æquales esse quadrato. Fingo quadratum à numeris quotquot libuerit, cum defectu tot vnitatum quot continet latus ipsius 16. esto à 2 N. − 4. ipse igitur quadratus erit 4 Q. + 16. − 16 N. hæc æquabuntur vnitatibus 16 − 1 Q. Communis adiiciatur vtrimque defectus, & à similibus auferantur similia, fient 5 Q. æquales 16 N. & fit 1 N. $\frac{16}{5}$ Erit igitur alter quadratorum $\frac{256}{25}$. alter verò $\frac{144}{25}$. & vtriusque summa est $\frac{400}{25}$ seu 16. & vterque quadratus est.

ΤΟΝ ἐπιταχθέντα τετράγωνον διελεῖν εἰς δύο τετραγώνους. ἐπιτετάχθω δὴ τὸν ις διελεῖν εἰς δύο τετραγώνους. καὶ τετάχθω ὁ πρῶτος δυνάμεως μιᾶς. δεήσει ἄρα μονάδας ις λείψει δυνάμεως μιᾶς ἴσας εἶναι τετραγώνῳ. πλάσσω τὸν τετράγωνον ἀπὸ ςς. ὅσων δή ποτε λείψει τοσούτων μ° ὅσων ἐστὶν ἡ τοῦ ις μ° πλευρά. ἔστω ςς β λείψει μ° δ. αὐτὸς ἄρα ὁ τετράγωνος ἔσται δυνάμεων δ μ° ις [λείψει ςς ις.] ταῦτα ἴσα μονάσι ις λείψει δυνάμεως μιᾶς. κοινὴ προσκείσθω ἡ λεῖψις, καὶ ἀπὸ ὁμοίων ὅμοια. δυνάμεις ἄρα ε ἴσαι ἀριθμοῖς ις. καὶ γίνεται ὁ ἀριθμὸς ις πέμπτων. ἔσται ὁ μὲν σνς εἰκοσοπέμπτων. ὁ δὲ ρμδ εἰκοσοπέμπτων, καὶ οἱ δύο συντεθέντες ποιοῦσι υ εἰκοσόπεμπτα, ἤτοι μονάδας ις. καί ἐστιν ἑκάτερος τετράγωνος.

DUKE UNIVERSITY SPECIAL COLLECTIONS LIBRARY

Pierre de Fermat's "Last Theorem" as reproduced in an edition of Diophantus' *Arithmetica* published by Fermat's son Samuel. The original copy of Diophantus with Fermat's handwritten note has never been found.

mat, who was Second Consul in the town of Beaumont-de-Lomagne, and of Claire de Long, the daughter of a family of parliamentary judges. The young Fermat was born in August, 1601 (baptized August 20 in Beaumont-de-Lomagne), and was raised by his parents to be a magistrate. He went to school in Toulouse, and was installed in the same city as Commissioner of Requests at the age of thirty. He married Louise Long, his mother's cousin, that same year, 1631. Pierre and Louise had three sons and two daughters. One of their sons, Clement Samuel, became his father's scientific executor and published his father's works after his death. In fact, it is the book containing Fermat's work, published by his son, that has come down to us and from which we know his famous Last Theorem. Clement Samuel de Fermat recognized the importance of the theorem scribbled in the margin and added it to the translation of the ancient work he republished.

Fermat's life is often described as quiet, stable, and uneventful. He did his work with dignity and honesty, and in 1648 was promoted to the important position of the King's Councillorship in the local Parliament of Toulouse, a title he held for seventeen years until his death in 1665. Considering the great work Fermat did for the Crown, a lifetime of devoted, able, and conscientious service, many historians are puzzled that he had the time and the mental energy to do first-rate mathematics—and volumes of it. One French expert suggested that Fermat's official work was actually an asset to his mathematic studies, since French parliamentary councilors were expected to minimize their unofficial contacts in order to avoid the temptations of bribery and other corruption. Since Fermat certainly required a diversion from his hard work, and since he had to limit his social life, mathematics probably offered a much-needed break. And the ideas of calculus were far from Fermat's only achievement. Fermat brought us number theory. An important element in number theory is the concept of a prime number.

Prime Numbers

The numbers one, two, and three are prime numbers. The number four is not prime because it is the product of two and two: 2 x 2 = 4. The number five is prime. The number six is not prime since, like four, it is the product of two numbers: 2 x 3 = 6. Seven is prime, eight is not (2 x 2 x 2=8), nine is not (3 x 3=9), and ten is not (2 x 5=10). But eleven again is a prime number since there are no integers (other than eleven itself and one), which can be multiplied together to give us 11. And we can continue this way: 12 is not prime, 13 is, 14 is not, 15 is not, 16 is not, 17 is prime, and so on. There is no apparent structure here, such as every fourth number is not a prime, or even any more complicated pattern. The concept has mystified human beings since early antiquity. Prime numbers are the essential elements in number theory, and the lack of easily-seen structure tends to make number theory seem un-unified as a field, and its problems isolated, difficult to solve, and without clear implications to other fields of mathematics. In the words of Barry Mazur: "Number theory produces, without effort, innumerable problems which have a sweet, innocent air about them, tempting flowers; and yet . . . number theory swarms with bugs, waiting to bite the tempted flower-lovers who, once bitten, are inspired to excesses of effort!"[2]

A Famous Note on the Margin

Fermat was smitten by the charm of numbers. In them he found beauty and meaning. He came up with a number of theorems in number theory, one of which was that every number of the form $2^{2^n}+1$ (two raised to the power two raised to the power n, plus one) is a prime number. Later, it was discovered that the theorem was false when a number of this form was found not to be a prime.

Among Fermat's cherished Latin translations of ancient texts was a book called the *Arithmetica*, written by the Greek mathematician Diophantus, who lived in Alexandria in the third century A.D. Around 1637, Fermat wrote in Latin in the margin of his Diophantus, next to a problem on breaking down a squared number into two squares:

> On the other hand, it is impossible to separate a cube into two cubes, or a biquadrate into two biquadrates, or generally any power except a square into two powers with the same exponent. I have discovered a truly marvelous proof of this, which, however, the margin is not large enough to contain.

This mysterious statement kept generations of mathematicians busy trying to supply the "truly marvelous proof" Fermat claimed to have possessed. The statement itself, that while a square of a whole number could be broken down into two other squares of whole numbers (for example, five squared, which is twenty-five, equals the sum of four squared (sixteen) and three squared [nine]), but that the same cannot be done with cubes or any higher powers, looked deceptively simple.

All of Fermat's other theorems were either proved or disproved by the early 1800s. This seemingly simple statement remained unsettled, and therefore was given the name "Fermat's Last Theorem." Was it indeed true? Even in our own century, computers were stymied in attempts to verify that the theorem was true. Computers could verify the theorem for very large numbers, but they couldn't help for *all* numbers. The theorem could be tried on billions of numbers, and there still would be infinitely many—and infinitely many exponents—to check. To establish Fermat's Last Theorem, a mathematical proof was required. Awards were offered in the 1800s by the French and German scientific academies to anyone who would come up with a proof, and every year thousands of mathematicians and amateurs, along with cranks, sent "proofs" to mathematical journals and judging committees—always coming up empty-handed.

July–August, 1993—A Fatal Flaw is Discovered

Mathematicians were cautiously optimistic when Wiles stepped down from the podium that Wednesday in June. Finally, the 350-year-old mystery seemed to have been solved. Wiles' lengthy proof, using complicated mathematical notions and theories which were not known during the time of Fermat or indeed until the twentieth century, needed to be validated by independent experts. The proof was sent to a number of leading mathematicians. Perhaps seven years of working alone in the seclusion of his attic had finally paid off for Wiles. But the optimism was short-lived. Within weeks, a hole was discovered in Wiles' logic. He tried to patch it, but the gap would simply not go away. Princeton mathematician Peter Sarnak, a close friend of Andrew Wiles, watched him agonize every day over the proof he had told the entire world he possessed only two months earlier in Cambridge. "It's as if Andrew was trying to lay an over-sized carpet on the floor of a room," Sarnak explained. "He'd pull it out, and the carpet would fit perfectly in one side of the room, but across the room it would be up against the wall, so he would go there and pull it down . . . and then it would pop up in another place. Whether or not the carpet had the right size for the room was not something he was able to determine." Wiles withdrew into his attic. The reporters from the *New York Times* and the rest of the media left him to his lonely task. As time went by without a proof, mathematicians and the public in general began to wonder whether Fermat's theorem was at all true. The marvelous proof Professor Wiles convinced the world he possessed became no more real than Fermat's own "truly marvelous proof which the margin is unfortunately too small to contain."

Between the Tigris and Euphrates Rivers, Circa 2000 *B.C.*

The story of Fermat's Last Theorem is much, much older than Fermat himself. It's even older than Diophantus, whose work Fermat was trying to generalize. The origins of this simple-looking yet profound theorem are as old as human civilization itself. They are rooted in the Bronze Age culture that developed in the Fertile Crescent between the Tigris and Euphrates rivers of ancient Babylon (an area within today's Iraq). And while Fermat's Last Theorem is an abstract statement with no applications in science, engineering, mathematics—not even in number theory, its own niche within mathematics—the roots of this theorem are grounded in the everyday life of the people of Mesopotamia of 2000 B.C. .

The era from 2000 B.C. to 600 B.C. in the Mesopotamian valley is considered the Babylonian Era. This time saw remarkable cultural developments, including writing, the use of the wheel, and metal works. A system of canals was used for irrigating large tracts of land between the two rivers. As civilization flourished in the fertile valley of Babylon, the ancient people who inhabited these planes learned to trade and to build prosperous cities such as Babylon and Ur (where Abraham was born). Even earlier, by the end of the fourth millennium B.C., a primitive form of writing had already developed in both the Mesopotamian and the Nile river valleys. In Mesopotamia, clay was abundant and wedge-shaped marks were impressed with a stylus on soft clay tablets. These tablets were then baked in ovens or left to harden in the sun. This form of writing is called cuneiform, a word derived from the Latin word *cuneus*, meaning wedge. The cuneiform constitutes the first known writing the world has ever seen. Commerce and construction in Babylon and in ancient Egypt brought the need for accurate measurements. The early scientists of these Bronze Age societies learned to estimate the ratio between the circumference and the diameter of a circle, which gave them a number close to what we call today pi. The people who built the giant Ziggurat, the biblical Tower of Babel, and the Hanging Gardens, one of the Seven Wonder of the Ancient World, needed a way to compute areas and volumes.

Wealth Is a Squared Quantity

A sophisticated number system was developed using base sixty, and Babylonian engineers and builders were able to compute the quantities required in their everyday professional lives. Squares of numbers appear naturally in life, although it doesn't seem so at first glance. Squares of numbers can be viewed as representing wealth. A farmer's prosperity is dependent on the amount of crops he is able to produce. These crops, in turn, depend on the *area* that is available to the farmer. The area is a product of the length and the width of the field, and this is where squares come in. A field that has length and width equal to a has area equal to a-squared. In this sense, therefore, wealth is a squared quantity.

The Babylonians wanted to know when such squares of whole numbers could be partitioned into other squares of whole numbers. A farmer who owned one field of twenty-five square units of land could swap it for two square fields: one measuring sixteen squared units and the other nine squared units. So a field five units by five units was equivalent to two fields, one four by four and the other three by three. This was important information for the solution of a practical problem. Today we would write this relationship in the form of an equation: $5^2 = 3^2 + 4^2$. And triples of such integers, here 3, 4, and 5, whose squares satisfy this relation, are called *Pythagorean triples*—even though they were known to the Babylonians over one thousand years before the time of the famous Greek mathematician, Pythagoras, after whom they are named. We know all this from an unusual clay tablet dated to around 1900 B.C.

"Plimpton 322"

The Babylonians were obsessed with tables. And the abundance of clay and the cuneiform writing technology they possessed allowed them to create many of them. Because of the clay tablets' durability, many of them survive today. From one location alone, the site of ancient Nippur, over 50,000 tablets were recovered and are now in the collections at the museums of Yale, Columbia, and the University of Pennsylvania, among others. Many of these tablets are in the basements of the museums, gathering dust, lying there unread and undeciphered.

One tablet that was deciphered is remarkable. This tablet, in the museum of Columbia University, is called Plimpton 322. All it contains are 15 triples of numbers. Each one of the triples has the property that the first number is a square and is the sum of the other two, each being itself a squared number—the table contains fifteen Pythagorean triples.[3] The numbers $25 = 16 + 9$, given earlier, form a Pythagorean triple. Another Pythagorean triple on Plimpton 322 is $169 = 144 + 25$ ($13^2 = 12^2 + 5^2$). Not all scholars agree on the reason for the ancient Babylonians' interest in these numbers. One theory is that the interest was solely for practical purposes, and the fact that they used a number system with base sixty and therefore preferred integers to fractions supports this need to solve practical problems with nice, whole square numbers. But other experts think that an inherent interest in numbers themselves may also have been a motivator for the Babylonians' interest in square numbers. It seems that, whatever the motive, Plimpton 322 may have served as a tool for teaching students to solve problems where the numbers are perfect squares.

COLUMBIA UNIVERSITY, RARE BOOK AND MANUSCRIPT LIBRARY

The Babylonians' approach was not to develop a general theory for solving such problems, but rather to provide tables listing triples of numbers and—apparently—to teach pupils how to read and use these tables.

An Ancient Society of Number-Worshippers Sworn to Secrecy

Pythagoras was born on the Greek island of Samos around 580 B.C. He travelled extensively throughout the ancient world and visited Babylon, Egypt, and possibly even India. In his travels, especially in Babylon, Pythagoras came in contact with mathematicians and likely became aware of their studies of numbers now named after him—the Pythagorean triples, which Babylonian scientists and mathematicians had known about for over 1500 years. Pythagoras came in contact with the builders of magnificent works of art and architecture, and the mathematical aspects of these wonders could not escape him. Pythagoras was also exposed in his travels to religious and philosophical ideas of the East.

When Pythagoras returned to Greece he left the island of Samos and moved to Crotona, on the Italian "boot." It is interesting to note that Pythagoras certainly saw most of the Seven Wonders of the Ancient World. One of these wonders, the Temple of Hera, is right where Pythagoras was born on Samos. Today, the ruins of the magnificent temple—only one standing column remains from among hundreds—are a short walk away from the modern town of Pythagorion, named in honor of the island's illustrious son. Just across the strait and a few miles to the north, in modern-day Turkey, lie another of the Wonders in the remains of ancient Ephesus. The Colossus of Rhodes is nearby, to the south of Samos; the Pyramids and the Sphynx are in Egypt and Pythagoras saw those; and in Babylon he must have seen the Hanging Gardens.

The Italian boot, including Crotona where Pythagoras settled, as well as much of the rest of southern Italy, were at that time part of the Greek world—Magna Graecia. This "greater Greece" included settlements all over the eastern Mediterranean, including Alexandria in Egypt with its large ethnic Greek population—descendants of which remained there through the early 1900s. Not far from Crotona were caves for oracles just like the Oracle of Delphi, who was believed to foretell fortunes and futures of people and nations.

"Number Is Everything"

In the barren, stark surroundings of the tip of Italy, Pythagoras founded a secret society dedicated to the study of numbers. The society, whose members became known collectively as the Pythagoreans, is believed to have developed a substantial body of mathematical knowledge—all in complete secrecy. The Pythagoreans are believed to have followed a philosophy summarized by their motto that "number is everything." They worshipped numbers and believed them to have magical properties. An object of interest to them was a "perfect" number. One of the definitions of a perfect number—a concept that continued to be pursued in the Middle Ages, and appears in mystical systems such as the Jewish Kabbalah—is a number that is the sum of its multiplicative factors. The best and simplest example of a perfect number is the number six. Six is the product of three and two and one. These are the multiplicative factors of this number, and we have: $6 = 3 \times 2 \times 1$. But note also that if you *add* the same factors you will again get the number six: $6 = 3 + 2 + 1$. In that sense, six is "perfect." Another perfect number is 28, since the numbers that can divide 28 (without remainder) are 1, 2, 4, 7, and 14, and we note that also: $1 + 2 + 4 + 7 + 14 = 28$.

The Pythagoreans followed an ascetic lifestyle, and were strict vegetarians. But they did not eat beans, thinking they resembled testicles. Their preoccupations with number were very much in the spirit of a religion, and their strict vegetarianism originated in religious beliefs. While no documents survive dating to the time of Pythagoras, there is a large body of later literature about the master and his followers, and Pythagoras himself is considered one of the greatest mathematicians of

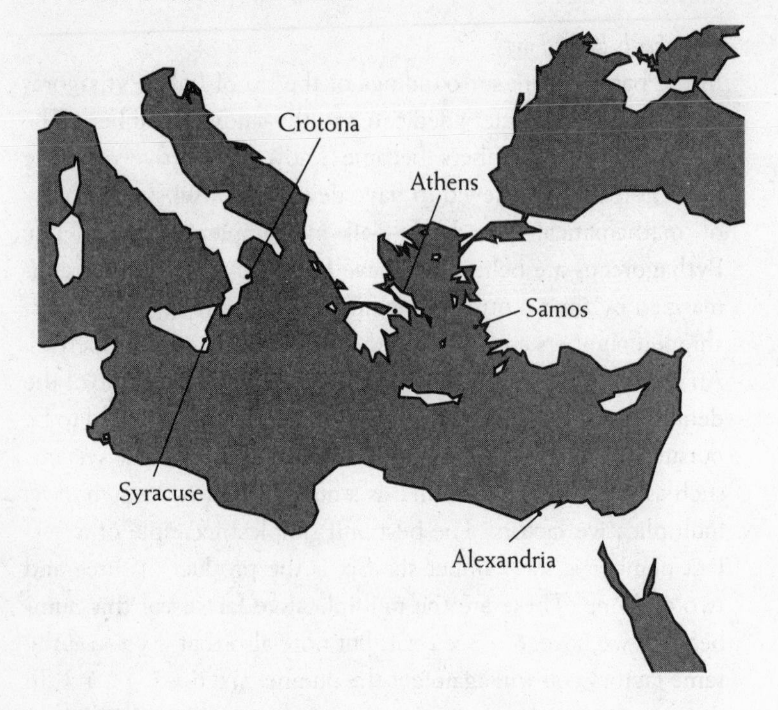

antiquity. To him is attributed the discovery of the Pythagorean Theorem concerning the squares of the sides of a right triangle, which has strong bearing on Pythagorean triples and, ultimately, on Fermat's Last Theorem two thousand years later.

The Square of the Hypotenuse Is Equal to the Sum of the Squares of the Other Two Sides . . .

The theorem itself originated in Babylon, since the Babylonians clearly understood "Pythagorean" triples. The Pythagoreans, however, are credited with setting the problem in geometric

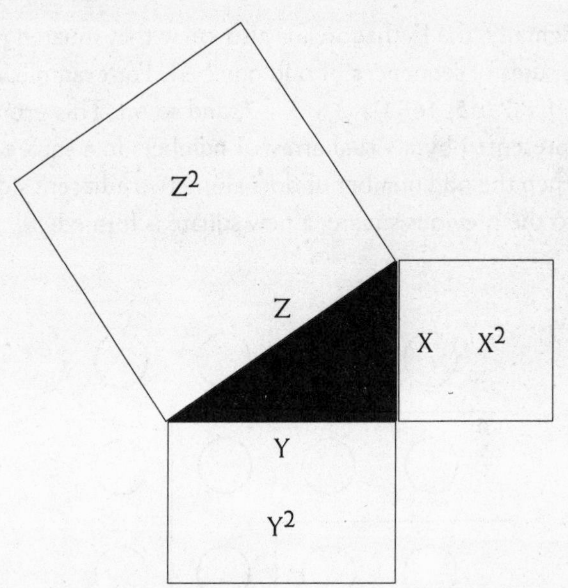

terms, and thus generalizing them away from strictly the natural numbers (positive integers without zero). The Pythagorean Theorem says that the square of the hypotenuse of a right-triangle is equal to the sum of the squares of the two remaining sides of the triangle, as shown above.

When the hypotenuse is an integer (such as 5, whose square is 25), the general Pythagorean solution in terms of the sum of two squares will be the integers four (whose square is sixteen) and three (whose square is nine). So the Pythagorean Theorem, when applied to integers (whole numbers such as 1, 2, 3, . . .) gives us the Pythagorean triples which were known a millennium earlier in Babylon.

Incidentally, the Pythagoreans also knew that squared numbers are sums of sequences of odd numbers. For example, 4 = 1 + 3; 9 = 1 + 3 + 5; 16 = 1 + 3 + 5 + 7, and so on. This property they represented by a visual array of numbers in a square pattern. When the odd number of dots along two adjacent sides is added to the previous square, a new square is formed:

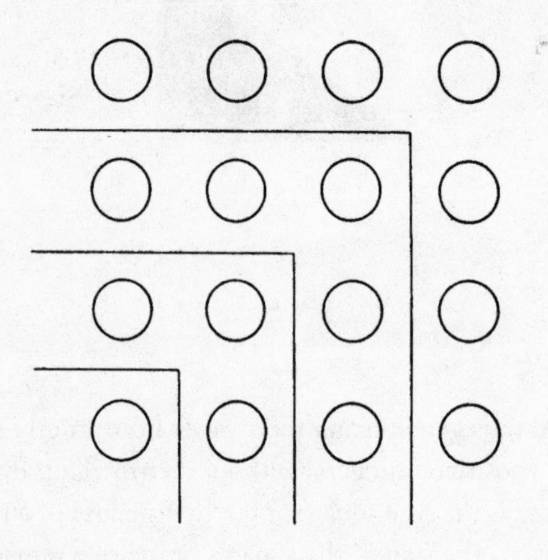

Whole Numbers, Fractions, and What Else?

But the Pythagoreans knew a lot more than whole numbers and fractions (numbers such as 1/2, 1/3, 5/8, 147/1769, etc.), which were known in antiquity both in Babylon and in Egypt. The Pythagoreans are the ones who discovered the *irrational* numbers—that is, numbers that cannot be written as fractions but have to be written as unending, non-repeating decimals. One such example is the number pi (3.141592654 . . .), the ratio of the circumference of a circle to its diameter. The number pi is unending; it would take forever to write it down completely since it has infinitely many (that is, neverending) distinct digits. To write it down, we simply say: pi (π). Or, we can write it to any finite number of decimals, such as 3.14, 3.1415, or so. Computers were used in our century to compute and write down pi to a million or more digits, but this is rarely necessary. Pi was known to within various approximations to the Babylonians and Egyptians of the second millennium B.C. They took it as roughly three, and it arose naturally as a consequence of the discovery of the wheel. Pi also arose in various measurements of a pyramid. Pi is even alluded to in the Old Testament: in Kings I, 7:23, we read about a circular wall being constructed. From the given number of units for the circumference and the diameter, we can conclude that the ancient Israelites took pi to be roughly three.

The Pythagoreans discovered that the square root of two was an irrational number. From an application of the Pythagorean Theorem to a right triangle with two sides both equal to one unit, the Pythagoreans obtained that the hypotenuse was a strange number: the square root of two. They could tell that this number was not an integer, nor even a fraction, a ratio of two integers. This was a number with an unending decimal representation which did not repeat itself. As is the case with pi, to write down the exact number that is the square root of two (1.414213562 . . .) would take forever, since there are infinitely many digits, forming a unique sequence (rather than a repeating one such as 1.857142857142857142857142857 . . . , etc., which one could describe without having to actually write down every digit). Any number with a repeating decimal representation (here, the sequence 857142 repeats itself over and over again in the decimal part of this number), is a *rational* number, that is, a number that can also be written in the form *a/b*, that is, the ratio of two integers. In this example, the two integers are 13 and 7. The ratio 13/7 is equal to 1.857142857142857142857142 . . . , the pattern 857142 repeating itself forever.

The discovery of the irrationality of the square root of two surprised and shocked these diligent number admirers. They swore never to tell anyone outside their society. But word got out. And legend has it that Pythagoras himself killed by drowning the member who divulged to the world the secret of the existence of the strange, irrational numbers.

The numbers on the number line are of two distinct kinds: rational and irrational. Looked at together, they fill the entire line with no holes. The numbers are very, very close (infinitesimally close) to each other. The rational numbers are said to be everywhere dense within the real numbers. Any neighborhood, any tiny interval, around a rational number contains infinitely many of these irrational numbers. And vice versa, around every irrational number there are infinitely many rational ones. Both sets, the rational and the irrational numbers, are infinite. But the irrationals are so numerous that there are more of them than there are rational numbers. Their order of infinity is higher. This fact was shown in the 1800s by the mathematician Georg Cantor (1845–1918). At the time, few people believed Cantor. His arch-enemy Leopold Kronecker (1823–1891) taunted and ridiculed Cantor for his theories about how many rational and irrational numbers there are. Kronecker is known for his statement, "God made the integers, and all the rest is the work of man," meaning that he did not even believe that the irrational numbers, such as the square root of two, existed! (This, over two millennia after the Pythagoreans.) His antagonism is blamed for having prevented Cantor from obtaining a professorship at the prestigious University of Berlin, and ultimately for Cantor's frequent nervous breakdowns and his ending up at an asylum for the mentally infirm. Today, all mathematicians know that Cantor was right and that there are infinitely many more irrational numbers than rational ones, even though both sets are infinite. But did the ancient Greeks know that much?[4]

Between the rationals are irrational numbers

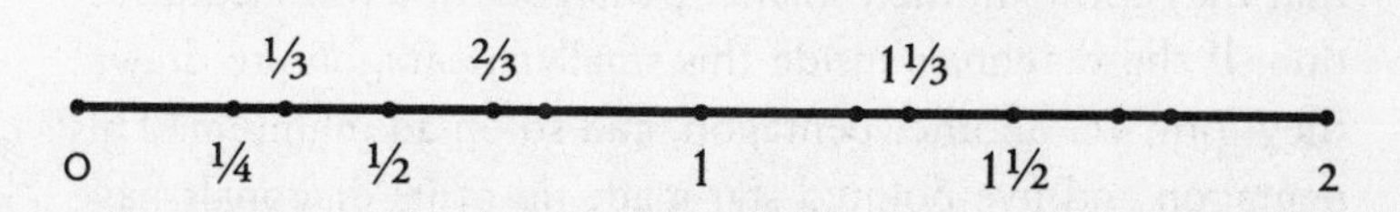

Rationals numbers are fractions

The Pythagorean Legacy

An important aspect of Pythagorean life, with its dietary rules and number worship and secret meetings and rituals, was the pursuit of philosophical and mathematical studies as a moral basis. It is believed that Pythagoras himself coined the words *philosophy*: love of wisdom; and *mathematics*: that which is learned. Pythagoras transformed the science of mathematics into a liberal form of education.

Pythagoras died around 500 B.C. and left no written records of his work. His center at Crotona was destroyed when a rival political group, the Sybaritics, surprised the members and murdered most of them. The rest dispersed about the Greek world around the Mediterranean, carrying with them their philosophy and number mysticism. Among those who learned the philosophy of mathematics from these refugees was Philolaos of Tarentum, who studied at the new center the Pythagoreans established in that city. Philolaos is the first Greek philosopher to have written down the history and theories of the Pythagorean order. It is from the book written by Philolaos that Plato learned of the Pythagorean philosophy of number, cosmology, and mysticism, about which he later wrote himself. The special symbol of the Pythagorean order was the five-pointed star embedded in a pentagon. The diagonals which form the five-pointed star intersect in such a way that they form another, smaller pentagon, in a reversed direction. If the diagonals inside this smaller pentagon are drawn, they form yet another pentagon, and so on ad infinitum. This pentagon and five-pointed star made up of its diagonals have some fascinating properties, which the Pythagoreans believed were mystical. A diagonal point divides a diagonal into two unequal parts. The ratio of the entire diagonal to the larger segment is exactly the same as the ratio of the larger segment to the smaller one. This same ratio exists in all smaller and smaller diagonals. This ratio is called the Golden Section. It is an irrational number equal to 1.618 . . . If you divide 1 by this number, you get exactly the decimal part without the 1. That is, you get 0.618. . . . As we will see later, the Golden Section appears in natural phenomena as well as proportions that the human eye perceives as beautiful. It appears as the limit of the ratio of the famous Fibonacci numbers we will soon encounter.

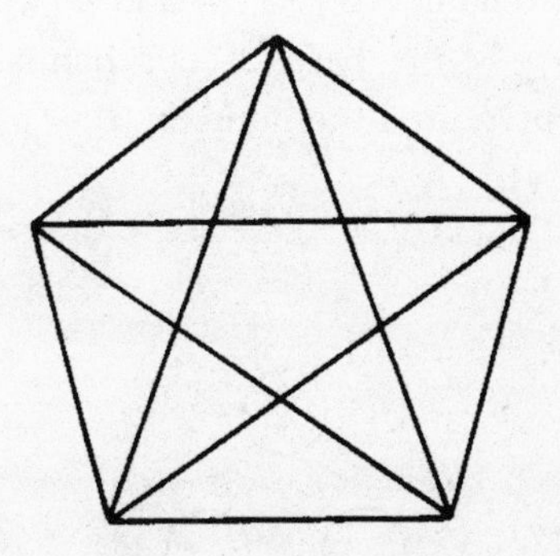

You can find the Golden Section by an interesting sequence of operations on a calculator. Do 1 + 1 =, then press 1/x, then +1=, then 1/x, then +1=, then 1/x and keep going. The number on your display should become 1.618... and 0.618... interchangeably, once you have done the repetitive set of operations enough times. This is the Golden Section. It is equal to the square root of five, minus 1, divided by 2. This is the way it is obtained geometrically from the Pythagorean Pentagon. Since this ratio never becomes a ratio of two integers, hence never a *rational* number, it proves that the square root of five is also an irrational number. We will see more of the Golden Section later.

The Pythagoreans discovered that harmony in music corresponded to simple ratios of numbers. According to Aristotle, the Pythagoreans believed that all of heaven was musical scale and numbers. Musical harmony and geometrical designs are what brought the Pythagoreans to their belief that "All is number." The Pythagoreans thought that the basic ratios in music involved only the numbers 1, 2, 3, and 4, whose sum is 10. And 10, in turn, is the base of our number system. The Pythagoreans represented the number 10 as a triangle, which they called *tetraktys*:[5]

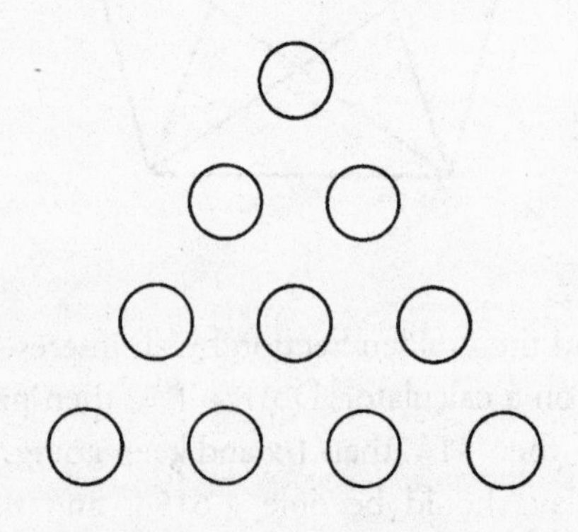

The Pythagoreans considered the tetraktys holy and even swore oaths by it. Incidentally, according to Aristotle as well as Ovid and other classical writers, the number ten was chosen as the base for the number system because people have ten fingers. Recall that the Babylonians, on the other hand, used a number system based on sixty. There are some remnants of other number systems even today. The French word for eighty (quatre-vingt, meaning "four twenties") is a relic of an archaic number system based on twenty.

The Ropes, The Nile, and the Birth of Geometry

Much of what we know about ancient Greek mathematics comes from the *Elements* of Euclid of Alexandria, who lived around 300 B.C. It is believed that the first two volumes of the *Elements* are all on the work of Pythagoras and his secret society. The mathematics of the ancient Greeks was done for its beauty and concerned abstract geometrical figures. The Greeks developed an entire theory of geometry and it is this theory, mostly unchanged, that is taught in schools today. In fact, the *Elements*, or what remains of it today, is considered the greatest textbook of all time.

Herodotus, the great Greek historian of antiquity, believed that geometry was developed in ancient Egypt of 3,000 B.C., long before the Greeks of Alexandria and elsewhere. He tells how the overflow of the Nile would destroy boundaries between fields in the river's fertile delta, and how this necessitated complicated surveying techniques. For the purpose of this work, the surveyors had to develop geometrical concepts and ideas. In his *Histories*, Herodotus writes:

> If the river carried away any portion of a man's lot, the king sent persons to examine and determine by measurement the exact extent of the loss. From this practice, I think, geometry first came to be known in Egypt, whence it passed into Greece.[6]

Geometry is the study of shapes and figures made of circles and straight lines and arcs and triangles and their intersections forming various angles. It stands to reason that such science would be essential for good surveying work. Egyptian geometers were indeed called "rope stretchers," since ropes were used for outlining *straight* lines necessary both in building temples and pyramids and in realigning boundaries between fields. But it is possible that the origins of geometry are even more ancient. Neolithic finds show examples of congruence and symmetry of design, and these may have been the precursors of Egyptian geometry, inherited centuries later by the ancient Greeks. The same concerns that the Babylonians had with areas of fields, leading to their need to understand square numbers and their relations, may have been shared by the ancient Egyptians, who were faced with the same agrarian quandaries as well as construction problems with their own pyramids. It is possible, therefore, that the ancient Egyptians also had a knowledge of Pythagorean triples. What the Greeks did with geometry, however, was to establish it as a pure mathematical endeavor. They postulated and proved theorems.

What is a Theorem?

The Greeks brought us the concept of a *theorem*. A theorem is a mathematical statement whose proof is given. The proof of a theorem is a rigorous justification of the veracity of the theorem in such a way that it cannot be disputed by anyone who follows the rules of logic, and who accepts a set of *axioms* put forth as the basis for the logic system. Euclid's axioms include the definition of a point, a line, and the statement that two parallel lines never meet. Following axioms and logical progressions, such as if A implies B and B implies C then A implies C, the ancient Greeks were able to prove many beautiful theorems about the geometry of triangles and circles and squares and octagons and hexagons and pentagons.

"Eureka! Eureka!"

The great Greek mathematicians Eudoxus (fifth century B.C.) and Archimedes (third century B.C.) extended such work on geometrical figures to the finding of areas using infinitesimal (meaning infinitely small) quantities. Eudoxus of Cnidus (408–355 B.C.) was a friend and student of Plato. He was too poor to live in the Academy in Athens, and so he lived in the cheaper harbor town of Piraeus, from which he commuted daily to Plato's Academy. While Plato himself was not a mathematician, he encouraged mathematical work, especially that of gifted students such as Eudoxus. Eudoxus traveled to Egypt and there, as well as in Greece, learned much geometry. He invented a "method of exhaustion," which he used to find areas of geometric figures by using infinitesimal quantities. Eudoxus would, for example, approximate the area of a circle by the sum of the areas of many small rectangles—whose areas are easy to calculate as the base times the height. This is essentially the method used today in integral calculus, and the modern limit arguments are not different from Eudoxus' "exhaustion" method.

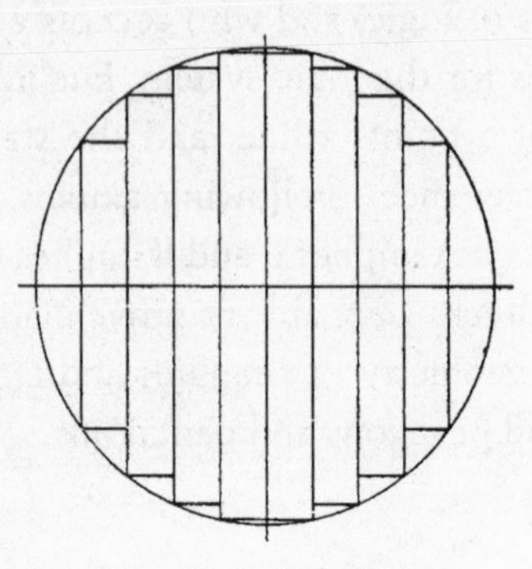

But the most brilliant mathematician of antiquity was undoubtedly Archimedes (287–212 B.C.), who lived in the city of Syracuse on the island of Sicily. Archimedes was the son of the astronomer Pheidias, and was related to Hieron II, the king of Syracuse. Like Eudoxus, Archimedes developed methods for finding areas and volumes, which were the forerunners of calculus. His work anticipated both integral calculus and differential calculus (there are two parts to calculus—Archimedes understood both of them). But while he was mostly interested in pure mathematics: numbers, geometry, areas of geometric figures, and so on, he is also known for his achievements in the applications of mathematics. A well-known story is the one about Archimedes' discovery of what we call today the first law of hydrostatics: the law that a submerged body loses from its weight the weight of the liquid it displaces. There was at that time a dishonest goldsmith in Syracuse, and King Hieron asked his mathematician friend to find a way to prove this. Archimedes started by studying the loss of weight of submerged bodies, using his own body in the experiments. He took a bath and made some measurements. When he discovered the law, he jumped out of the bath and ran naked through the streets of Syracuse shouting "Eureka, eureka!" ("I found it, I found it!").

Archimedes is also credited with the discovery of archimedes' screw, a device to raise water by turning a hand-operated crank. It is still in use by farmers around the world.

When the Roman General Marcellus attacked Syracuse in 214–212 B.C., Hieron again asked for the help of his illustrious relative. As the Roman fleet was approaching, Archimedes devised great catapults based on his study of levers, and the Syracusans were able to defend themselves well. But Marcellus regrouped his forces and some time later attacked from the rear and was able to take Syracuse by surprise. This time Archimedes was not even aware of the attack and was sitting quietly on the ground above the city drawing geometrical figures in the sand. A Roman soldier approached and stepped on the figures. Archimedes jumped, exclaiming: "Don't disturb my circles!" at which the soldier drew his sword and killed the 75-year-old mathematician. In his will, Archimedes had apparently requested that his gravestone bear a carving of the particular geometric figure he admired—a sphere inside a cylinder. The neglected grave was covered and the site lost, but the Roman orator Cicero found it many years later and restored it, and then the sands of time covered it again. In 1963, ground was broken for a new hotel near Syracuse, and there workers rediscovered Archimedes' tomb.

Archimedes' favorite theorem had to do with the sphere inside the cylinder and he wrote the theorem in a book called *The Method*. As with most ancient texts, it was assumed lost. In 1906, the Danish scholar J. L. Heiberg heard that in Constantinople there was a faded parchment manuscript with writings of a mathematical nature. He traveled to Constantinople and found the manuscript, consisting of 185 leaves of parchment. Scientific studies proved it was a tenth-century copy of Archimedes' book, over which Eastern Orthodox prayers had been added in the thirteenth century.

Alexandria, Greek Egypt, Circa A.D. 250

Around A.D. 250 a mathematician by the name of Diophantus lived in Alexandria. All that we know about the life of Diophantus is what is given in the following problem in a collection called the *Palatine Anthology*, written roughly a century after Diophantus' death.[7]

> Here you see the tomb containing the remains of Diophantus, it is remarkable: artfully it tells the measures of his life. The sixth part of his life God granted him for his youth. After a twelfth more his cheeks were bearded. After an additional seventh he kindled the light of marriage, and in the fifth year he accepted a son. Alas, a dear but unfortunate child, half of his father he was when chill Fate took him. He consoled his grief in the remaining four years of his life. By this devise of numbers, tell us the extent of his life.

(If you solve the implied equation, you will find that the answer is 84.)

It is not certain when Diophantus lived. We can date his period based only on two interesting facts. First, he quotes in his writings Hypsicles, whom we know lived around 150 B.C. And second, Diophantus himself is quoted by Theon of Alexandria. Theon's time is dated well by the solar eclipse which occurred on June 16, A.D. 364. So Diophantus certainly lived before A.D. 364 but after 150 B.C. Scholars, somewhat arbitrarily, place him at about A.D. 250.

Diophantus wrote the *Arithmetica*, which developed algebraic concepts and gave rise to a certain type of equation. These are the Diophantine equations, used in mathematics today. He wrote fifteen volumes, only six of which came down to us. The rest were lost in the fire that destroyed the great library of Alexandria, the most monumental collection of books in antiquity. The volumes that survived were among the last Greek texts to be translated. The first known Latin translation was published in 1575. But the copy Fermat had was the one translated by Claude Bachet in 1621. It was Diophantus' Problem 8 in Volume II, asking for a way of dividing a given square number into the sum of two squares—the Pythagorean problem whose solution was known to the Babylonians two thousand years earlier—that inspired Fermat to write his famous Last Theorem in the margin. The mathematical achievements of Diophantus and his contemporaries were the final glory of the ancient Greeks.

Arabian Nights

While Europe was busy fighting little feudal wars of the vassals of one king or prince against another, surviving the Great Plague, and going on costly and often deadly expeditions called the Crusades, the Arabs ruled a flourishing empire from the Middle East to the Iberian Peninsula. Among their great achievements in medicine, astronomy, and the arts, the Arabs developed algebra. In A.D. 632, the prophet Mohammed established an Islamic state centered at Mecca, which remains the religious center of Islam. Shortly afterwards, his forces attacked the Byzantine Empire, an offensive which continued after Mohammed's death in Medina that same year. Within a few years, Damascus, Jerusalem, and much of Mesopotamia fell to the forces of Islam, and by A.D. 641 so did Alexandria—the mathematical center of the world. By A.D. 750, these wars as well as the ones among the Moslems themselves subsided and the Arabs of Morocco and the west were reconciled with the eastern Arabs centered in Baghdad.

Baghdad became a center of mathematics. The Arabs absorbed mathematical ideas as well as discoveries in astronomy and other sciences from the inhabitants of the areas they overcame. Scholars from Iran, Syria, and Alexandria were called to Baghdad. During the reign of the caliph Al Mamun in the early 800s, the *Arabian Nights* was written and many Greek works, including Euclid's *Elements,* were translated into Arabic. The caliph established a House of Wisdom in Baghdad, and one of its members was Mohammed Ibn Musa Al-Khowarizmi. Like Euclid, Al-Khowarizmi was to become world-renowned. Borrowing Hindu ideas and symbols for numerals, as well as Mesopotamian concepts and Euclid's geometrical thought, Al-Khowarizmi wrote books on arithmetic and algebra. The word "algorithm" is derived from Al-Khowarizmi. And the word "algebra" is derived from the first words in the title of Al-

Khowarizmi's most well-known book: *Al Jabr Wa'l Muqabalah*. It was from this book that Europe was later to learn the branch of mathematics called algebra. While algebraic ideas are in the root of Diophantus' *Arithmetica*, the *Al Jabr* is more closely related to the algebra of today. The book is concerned with straightforward solutions of equations of first and second degree. In Arabic, the name of the book means "restoration by transposing terms from one side of an equation to the other"—the way first-order equations are solved today.

Algebra and geometry are related, as are all branches of mathematics. One field that links these two together is algebraic geometry, developed in our century. It is the linkage of branches of mathematics, and areas that lie within different branches and connect them, that would pave the way to Wiles' work on Fermat's problem centuries later.

The Medieval Merchant and the Golden Section

The Arabs were interested in a problem that was very closely related to the Diophantine question of finding Pythagorean triples. The problem was to find Pythagorean triples giving an area of a right-triangle that is also an integer. Hundreds of years later, this problem found itself as the basis for the book *Liber Quadratorum*, written in 1225 by Leonardo of Pisa (1180–1250). Leonardo was better known as Fibonacci (which means "son of Bonaccio"). Fibonacci was an international merchant born in Pisa. He also lived in North Africa and Constantinople, and he traveled extensively throughout his life and visited Provence, Sicily, Syria, Egypt, and many other areas in the Mediterranean. His travels and his relations with the elite of Mediterranean society of the time brought him into contact with Arab mathematical ideas, as well as Greek and Roman culture. When the emperor Frederick II came to Pisa, Fibonacci was introduced to the emperor's court and became a member of the imperial entourage.

In addition to *Liber Quadratorum*, Fibonacci is known for another book he wrote at that time, *Liber Abaci*. A problem about Pythagorean triangles in Fibonacci's book also appears in a Byzantine manuscript of the eleventh century now in the Old Palace library in Istanbul. It could be a coincidence; on the other hand, Fibonacci might have seen that same book in Constantinople during his travels.

Fibonacci is best known for the sequence of numbers named after him, the Fibonacci Numbers. These numbers originate in the following problem in the *Liber Abaci*:

> How many pairs of rabbits will be produced in a year, beginning with a single pair, if in every month each pair bears a new pair which becomes productive from the second month on?

The Fibonacci sequence, which is derived from this problem, is one where each term after the first is obtained by adding together the two numbers that precede it. The sequence is: 1, 1, 2, 3, 5, 8, 13, 21, 34, 55, 89, 144, . . .

This sequence (which is taken to continue beyond the 12 months of the problem) has unexpectedly significant properties. Amazingly, the ratio of two successive numbers in the sequence tends to the Golden Section. The ratios are: 1/1, 1/2, 2/3, 3/5, 5/8, 8/13, 13/21, 21/34, 34/55, 55/89, 89/144, etc. Note that these numbers get closer and closer to (sqr5 - 1)/2.

This is the Golden Section. It can also be obtained using a calculator by repeating the operation 1/1 + 1/1 + 1/. . . . as described earlier. Recall that the reciprocal (1/x) of the Golden Section gives the same number, less 1. The Fibonacci sequence appears everywhere in nature. Leaves on a branch grow at distances from one another that correspond to the Fibonacci sequence. The Fibonacci numbers occur in flowers. In most flowers, the number of petals is one of: 3, 5, 8, 13, 21, 34, 55, or 89. Lilies have three petals, buttercups five, delphiniums often eight, marigolds thirteen, asters twenty-one, daisies usually thirty-four or fifty-five or eighty-nine.

The Fibonacci numbers appear in sunflowers too. The little florets that become seeds in the head of the sunflower are arranged in two sets of spirals: one winding in a clockwise direction and the other counter-clockwise. The number of spirals in the clockwise orientation is often thirty-four and the counter-clockwise fifty-five. Sometimes the numbers are fifty-five and eighty-nine, and sometimes even eighty-nine and a hundred and forty-four. All are consecutive Fibonacci numbers (whose ratio approaches the golden section). Ian Stewart argues in *Nature's Numbers* that when spirals are developed, the angles between them are 137.5 degrees, which is 360 degrees multiplied by one minus the golden ratio, and they also give rise to two successive Fibonacci numbers for the number of clockwise and counter-clockwise spirals, as shown below.[8]

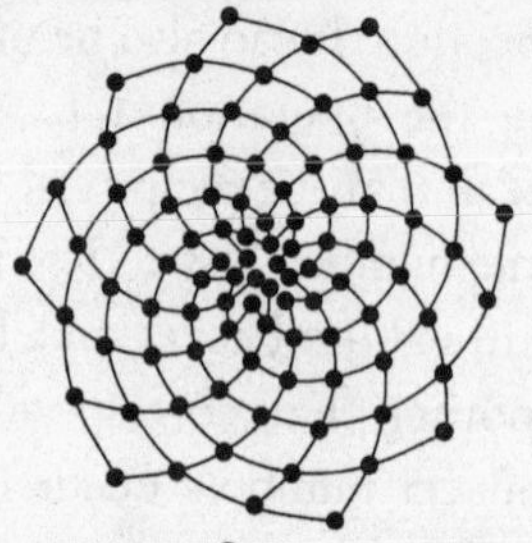

If a rectangle is drawn with sides in the Golden Section ratio to each other, then the rectangle can be divided into a square and another rectangle. This second rectangle is similar to the large one in that it, too, has ratio of sides equal to the Golden Section. The smaller rectangle can now be divided into a square and a remaining rectangle, also in the Golden ratio . . . and so on. A spiral through successive vertices of the sequence of rectangles that can be drawn is one that appears often in shells, in the arrangement of sunflower florets as mentioned, and in the arrangement of leaves on a branch.

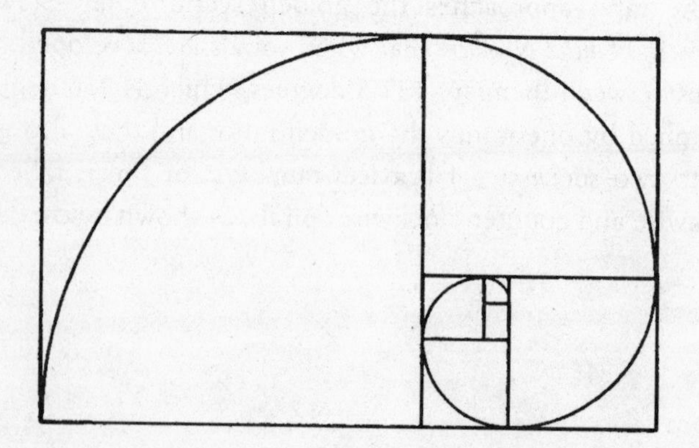

COURTESY GREEK NATIONAL TOURIST OFFICE

The Parthenon, Athens, Greece.

The rectangle has appealing proportions. The Golden Section appears not only in nature but also in art as the classic ideal of beauty. There is something divine about the sequence, and in fact the Fibonacci Society, which is active today, is headed by a priest and centered at St. Mary's College in California. The Society is dedicated to the pursuit of examples of the Golden Section and Fibonacci Numbers in nature, art, and architecture, with the belief that the ratio is a gift of God to the world. As the ideal of beauty, the Golden Section appears in such places as the Athenian Parthenon. The ratio of the height of the Parthenon to its length is the Golden Section.

The great Pyramid at Giza, built many hundreds of years before the Greek Parthenon, has ratio of height of a face to half the side of the base also in the Golden Section. The Egyptian Rhind Papyrus refers to a "sacred ratio." Ancient statues as well as renaissance paintings display the proportions equal to the Golden Section, the Divine Ratio.

The Golden Section has been searched for as the ideal of beauty beyond flowers or architecture. In a letter to the Fibonacci Society some years ago, a member described how someone looking to find the Golden Section asked several couples to perform an experiment. The husband was asked to measure the height of his wife's navel and divide it by his wife's height. The writer claimed that in all couples the ratio was close to 0.618.

The Cossists

Mathematics entered Medieval Europe through Fibonacci's works and from Spain, then part of the Arab world, with the work of Al Khowarizmi. The main idea of algebra in those days was to solve equations for an unknown quantity. Today, we call the unknown quantity "x," and try to solve an equation for whatever value "x" may have. An example of the simplest equation is: $x - 5 = 0$. Here, we use simple math operations to find the value of "x." If we add 5 to both sides of the equation we get on the left side $x - 5 + 5$, and on the right side we get $0 + 5$. So the left side is "x" and the right side is 5. That is, $x = 5$. The Arabs of Al Khowarizmi's day called the unknown quantity "thing." The word "thing" in Arabic is *shai*. And so they solved equations for the unknown shai, as done above with "x."

When these ideas were imported to Europe, the Arabic *shai* was translated into Latin. In Latin, "thing" is *res*, and in Italian it is *cosa*. Since the first European algebraists were Italian, the word *cosa* was attached to them. Since they were concerned with solving equations for an unknown *cosa*, they became known as the Cossists.[9]

As in Babylon three and a half millennia earlier, mathematics in the Middle Ages and the early Renaissance was mainly used as an aid in commerce. Mercantile society of that time was increasingly concerned with problems of trade, exchange rates, profits, costs, and these could sometimes be cast as mathematical problems requiring the solution of some equation. The cossists were people such as Luca Pacioli (1445–1514), Geronimo Cardano (1501–1576), Niccolo Tartaglia (1500–1557), and others who competed with each other as problem-solvers in the service of merchants and traders. These mathematicians used the solution of more abstract problems as a form of advertising. Since they had to compete for clients, they would also spend time and effort solving these more difficult problems, such as cubic equations (equations where the unknown quantity "cosa," or our "x," is in the third power, x^3), so that they could publish the results and become ever more sought-after for the solution of applied problems.

In the early 1500s, Tartaglia found a way to solve cubic equations and kept the method secret, so he could retain an edge over his competitors in the lucrative problem-solving market. After Tartaglia won a problem-solving competition against another mathematician, Cardano pressed him to reveal his secret of how he could solve these cubic equations.

Tartaglia revealed his method, on the condition that Cardano keep it a secret from the rest of the world. When Cardano later learned the same methods from another cossist, Scippione del Ferro (1456–1526), he immediately assumed that Tartaglia got his system from this person, and felt free to reveal the secret. Cardano then published the methodology for solving cubic equations in his 1545 book *Ars Magna*. Tartaglia felt betrayed and became furious with Cardano. In his last years, he spent much of his time vilifying his former friend, and he succeeded in diminishing Cardano's reputation.

The cossists were considered mathematicians of a lower level than the ancient Greeks. Their preoccupation with applied problems in pursuit of financial success, and their unconstructive fights among themselves, kept them from looking for beauty in mathematics and the pursuit of knowledge for its own sake. They did not develop an abstract, general theory of mathematics. For that, one needed to go back to the ancient Greeks. That is exactly what happened a century later.

Renaissance Quest for Ancient Knowledge

Thirteen hundred years had passed since Diophantus. The medieval world gave way to the Renaissance and the beginning of the modern age. Out of the darkness of the Middle Ages, Europe awoke with a thirst for knowledge. Many people turned their interest to classical works of the ancients. Whatever ancient books existed were translated into Latin—the language of the educated—in this revival of the search for knowledge and enlightenment. Claude Bachet, a French nobleman, was a translator with a great interest in mathematics. He obtained a copy of the Greek *Arithmetica* of Diophantus, translated it and published it as *Diophanti Alexandrini Arithmeticorum Libri Sex* in Paris in 1621. It was a copy of this book that found its way to Fermat.

Fermat's theorem says is that there are no possible Pythagorean triples for anything beyond squares. There are no triples of numbers, two adding up to the third, where the three numbers are perfect cubes of integers, or fourth-powers of integers, fifth, sixth, or any other powers. How could Fermat come up with such a theorem?

Squares, Cubes, and Higher Dimensions

A theorem is a statement with a proof. Fermat claimed to have had a "marvelous proof," but without seeing and validating the proof no one could call his statement a theorem. A statement may be very deep, very meaningful and important, but without the proof that it is indeed true, it must be called a conjecture, or sometimes a hypothesis. Once a conjecture is proven it then can be called a *theorem*, or a *lema* if it is a preliminary proven statement which then leads to a more profound theorem. Proven results that follow a theorem are called corollaries. And Fermat had a number of such statements. One such statement was that the number $2^{2^\wedge n}+1$ was always a prime number. This conjecture was not only not proven, hence not a theorem, it was actually proven to be *wrong*. This was done by the great Swiss mathematician Leonhard Euler (1707–1783) in the following century. So there was no reason to believe that the "last theorem" was true. It could be true, or it could be false. To prove that Fermat's Last Theorem was false all someone would have to do is to find a triple of integers, a, b, and c, and a power n, greater than 2, that satisfy the relation $a^n + b^n = c^n$. No one has ever found such a set of integers. (However, assuming that a solution exists was to be a key element in attempts to prove the theorem later.) And by the 1990s it was shown that no such integers exist for any n less than four million. But that did not mean that such numbers could not be found some day. The theorem had to be *proven* for *all* integers and *all* possible powers.

Fermat himself was able to prove his Last Theorem for $n=4$. He used an ingenious method he called the method of "infinite descent," to prove that no integers a, b, and c exist satisfying $a^4 + b^4 = c^4$. He also recognized that if a solution existed for any power n, then it would also exist for any multiple of n. One would therefore only have to consider prime numbers (greater than 2) as the exponents, that is, numbers that cannot be divided by any number other than 1 or themselves in integers. The first few prime numbers are 1, 2, 3, 5, 7, 11, 13, 17... None of these numbers can be divided by any number other than 1 or themselves and an integer result. An example of a number that is not a prime is 6, since 6 divided by 3 gives 2—an integer. Fermat was also able to prove his theorem for $n=3$. Leonhard Euler proved the case $n=3$ and $n=4$ independently of Fermat, and Peter G. L. Dirichlet in 1828 was able to prove the case $n=5$. The same case was proved by Adrien-Marie Legendre in 1830. Gabriel Lamé, and Henri Lebesgue who corrected him in 1840, were able to establish the case $n=7$. Thus, two hundred years after Fermat wrote his famous note in the margin of his Diophantus, his theorem was only proven correct for the exponents 3, 4, 5, 6, and 7. It would be a long way to infinity, which is what one had to do to prove the theorem for *any* exponent n. Clearly, what was needed was a general proof that would work for all exponents, however large they might be. Mathematicians were all looking for the elusive general proof, but unfortunately what they were finding were proofs for particular exponents only.

The Algorist

An algorist is a person who devises computing systems, or algorithms. One such person was the prolific Swiss mathematician Leonhard Euler, who was said to be able to calculate as naturally as other people breathed. But Euler was much more than a walking calculator. He was the most productive Swiss scientist of all times, and a mathematician who wrote so many volumes of work that the Swiss government established a special fund to collect all of his works. He is said to have produced mathematical papers between two calls for dinner at his large household.

Leonhard Euler was born in Basel on April 15, 1707. The following year, the family moved to the village of Riechen, where the father became the Calvinist pastor. When young Leonhard went to school, his father encouraged him to pursue the study of theology so that he would eventually take his place as the village pastor. But Euler showed much promise in mathematics and was tutored by Johannes Bernoulli, a well-known Swiss mathematician of the day. Daniel and Nicolaus Bernoulli, two younger members of the large Bernoulli family of mathematicians, became his good friends. These two convinced Leonhard's parents to allow him to pursue mathematics, since he would become a great mathematician. Leonhard, however, continued with the theology in addition to mathematics, and religious feelings and customs would be a part of his entire life.

Mathematicial and scientific research in Europe in those days did not take place primarily at universities, as it does today. The universities were more devoted to teaching and did not allow much time for other activities. Research in the eighteenth century was primarily done at royal academies. There, the monarch would support the leading scientists of the day in their pursuit of knowledge. Some of the knowledge was applied, and would help the government improve the standing of the entire nation. Other research was more "pure," that is, research for its own sake—for the advancement of human knowledge. The royals supported such research generously and the scientists working at the academies were able to lead comfortable lives.

When he finished his studies of mathematics, as well as theology and Hebrew, at the University of Basel, Euler applied for a professorship. Despite the great achievements he had already made, he was turned down. In the meantime, his two friends Daniel and Nicolaus were appointed as research mathematicians at the royal academy in St. Petersburg, Russia. The two stayed in touch with Leonhard and promised that, somehow, they would get him there too. One day, the two Bernoullis wrote an urgent letter to Euler telling him that there was an opening in the medical section of the St. Petersburg academy. Euler immediately put himself to work studying physiology and medicine at Basel. Medicine was not something that interested him, but he was desperate to have a job and hoped that this way he could join his two friends who had such excellent positions doing nothing but their own research in Russia.

Euler found mathematics in whatever he studied, medicine included. Studying ear physiology led him to a mathematical analysis of the propagation of waves. At any rate, soon an invitation came to St. Petersburg, and in 1827 he joined his two friends. However, on the death of Peter the Great's wife Catherine, there was chaos at the academy since she had been the great supporter of research. In the confusion, Leonhard Euler slipped out of the medical section and somehow got his name listed with the mathematical section, where he would rightfully belong. For six years he kept his head low to prevent the detection of his move, and he avoided all social interactions lest the deception be discovered. All through this period he worked continuously, producing volumes of top-rated mathematical work. In 1733 he was promoted to the leading mathematical position at the academy. Apparently Euler was a person who could work anywhere, and as his family was growing, he would often do his mathematics while holding a baby in one arm.

When Anna Ivanova, Peter the Great's niece, became empress of Russia, a period of terror began and Euler again hid himself in his work for ten years. During this period he was working on a difficult problem in astronomy for which a prize was offered in Paris. A number of mathematicians requested several months' leave from the academy to work on the problem. Euler solved it in three days. But the concentrated exertion took its toll and he became blind in his right eye.

Euler moved to Germany to be at the royal academy there, but did not get along with the Germans, who enjoyed long philosophical discussions not to his taste. Catherine the Great of Russia invited Euler back to the St. Petersburg academy and he was more than happy to come back. At that time, the philosopher Denis Diderot, an atheist, was visiting Catherine's court. The empress asked Euler to argue with Diderot about the existence of God. Diderot, in the meantime, was told that the famous mathematician had a proof of God's existence. Euler approached Diderot and said gravely: "Sir, $a + b/n = x$, hence God exists; reply!" Diderot, who knew nothing about mathematics, gave up and immediately returned to France.

During his second stay in Russia, Euler went blind in his second eye. He continued, however, to do mathematics with the help of his sons, who did the writing for him. Blindness increased his mental ability to do complicated calculations in his head. Euler continued to do mathematics for seventeen years and died while playing with his grandson in 1783. Much of the mathematical notation we use today is due to Euler. This includes the use of the letter i for the basic imaginary number, the square root of -1. Euler loved one mathematical formula, which he considered the most beautiful and put it above the gates of the Academy. The formula is:

$$e^{i\pi} + 1 = 0$$

This formula has 1 and 0, basic to our number system; it has the three mathematical operations: addition, multiplication, and exponentiation; and it has the two natural numbers pi and e; and it has i, the basis for the imaginary numbers. It is also visually appealing.

The Seven Bridges of Königsberg

Euler was such an incredible visionary in mathematics that his pioneering work on imaginary numbers (and what today is called complex analysis), was not his only innovation. He did pioneering work in a field which, in our century, would become indispensable in the work of mathematicians—and in attempts to solve the Fermat mystery. The field is topology, a visual theory of spatial configurations that can remain unchanged when transformed by continuous functions. It is the study of shapes and forms, some with intricate, unexpected geometry, which is extended to four, five, or higher dimensions beyond our normal three-dimensional world. We will visit this fascinating area again when we get to the modern approach to Fermat's problem, since topology—much as it seems unrelated to the Fermat equation—has great importance for understanding it.

Predating the development of topology, Euler's contribution to the field is the famous problem of the Seven Bridges of Königsberg. This is the puzzle that started the whole interest in topology. In Euler's time, seven bridges crossed the Pregel River in Königsberg. These are shown in the diagram below.

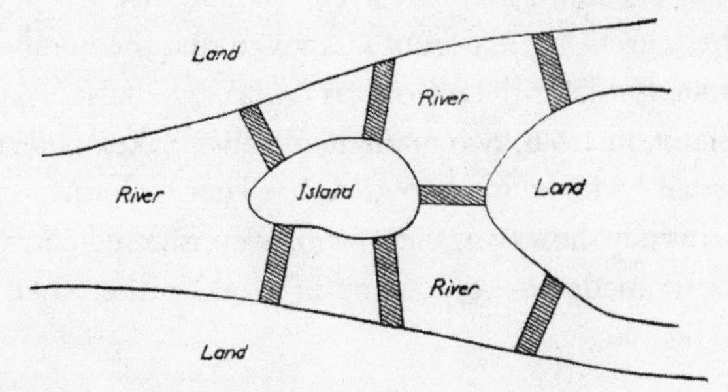

Euler asked whether or not it was possible to cross all seven bridges without passing twice on any bridge. It is impossible. Other problems, which were studied in modern times and were posed because of interest in the seven bridges problem, are the various map-coloring problems. A cartographer draws a map of the world. In this map, each country is colored differently, to distinguish it from its immediate neighbors. Any two countries or states that are separated completely from each other may be colored in the same exact color. The question is, what is the minimum number of colors required so that no two states that touch each other are in the same color? Of course, this is a general problem, not bounded by how the map of the world looks today. The question is really, given all possible configurations of maps on a plane, what is the minimum number of colors that can be used? Given boundaries between states in the former Yugoslavia or in the Middle East, with very unusual lines between political entities, this general problem becomes relevant in applications.

Mathematically, this is a topological problem. In October 1852, Francis Guthrie was coloring a map of England. He wondered what would be the minimum number of colors to be used for the counties. It occurred to him that the number should be four. In 1879 a proof was given that the number was indeed four, but later the proof was found to be false. Almost a century later, in 1976, two mathematicians, Haken and Appel, proved what had become known as the Four Color Map problem. To this day, however, their proof is considered controversial since it made use of computer work, rather than pure mathematical logic.

Gauss, The Great German Genius

An alleged error in Euler's proof for $n=3$ (that is, for cubes) was corrected by Carl Friedrich Gauss (1777–1855). While most of the renowned mathematicians of this time were French, Gauss, undoubtedly the greatest mathematician of the time—and arguably of all times—was unmistakably German. In fact, he never left Germany, even for a visit. Gauss was the grandson of a very poor peasant, and the son of a laborer in Brunswick. His father was harsh with him, but his mother protected and encouraged her son. Young Carl was also taken care of by his uncle Friedrich, the brother of Gauss' mother Dorothea. This uncle was wealthier than Carl's parents and made a reputation for himself in the field of weaving. When Carl was three years old, he once watched his uncle add up his accounts in a ledger. "Uncle Friedrich," he interrupted, "this calculation is wrong." The uncle was shocked. From that day on, the uncle did everything possible to contribute to the education and care of the young genius. Although Gauss showed incredible promise in school, his behavior sometimes left something to be desired. One day, the teacher punished young Gauss by telling him to stay in the classroom until he added up all the numbers from 1 to 100 while everyone else went to play outside. Two minutes later, the 10-year-old Gauss was outside playing with the rest of the class. The teacher came out furious. "Carl Friedrich!" the teacher called, "do you want a more severe punishment? I told you to stay inside until you have finished adding up all the numbers!" "But I have," he said, "here is the answer." Gauss handed the teacher a piece of paper with the right answer,

5,050, written on it. Apparently, Gauss figured out that he could write two rows of 101 numbers:

0	1	2	3..........	97	98	99	100
100	99	98	97..........	3	2	1	0

He noted that the sum of each column was 100, so there was nothing long to add up. Since there were 101 columns, the sum of all the numbers was 101 x 100=10,100. Now, either of the two rows had the sum he needed—all the numbers from 1 to 100. Since he needed only one of the two rows, the answer was half of 10,100, or 5,050. Very simple, he thought. The teacher, however, learned a lesson and never again assigned the young Gauss a math problem as punishment.

When he was fifteen, Gauss attended college at Brunswick, through the help of the Duke of Brunswick. The Duke later also supported the young mathematician in enrolling at the renowned university at Göttingen. There, on March 30, 1796, Gauss wrote the first page in his famous diary. The diary had only nineteen pages, but in these pages Gauss recorded 146 brief statements of important and powerful mathematical results he had derived. It was later discovered that almost every important mathematical idea published by any mathematician in the late eighteenth and in the nineteenth centuries had been preceded by one of the entries in Gauss' unpublished diary. The diary remained hidden until it was found in the possession of Gauss' grandson in Hamlin in 1898.

Gauss' results in number theory, which were shared with the mathematicians of his day by regular correspondence, were of great importance in all attempts by mathematicians to prove

Fermat's Last Theorem. Many of these results were contained in a book on number theory which Gauss published in Latin in 1801, when he was 24. The book, *Disquisitiones Arithmeticae*, was translated into French and published in Paris in 1807 and received much attention. It was recognized as the work of genius. Gauss dedicated it to his patron, the Duke of Brunswick.

Gauss was a distinguished scholar of classical languages as well. On entering college he was already a master of Latin, and his interest in philology precipitated a crisis in his career. Should he follow the study of languages or of mathematics? The turning point was March 30, 1796. From his diary, we know that on that day the young Gauss decided definitely to specialize in mathematics. In mathematics and statistics (where he is credited with the ingenious method of least squares for finding a line to fit a data set) he contributed to many areas, but he believed that number theory was the heart of all of mathematics.

But why did the world's greatest mathematical genius never try to prove Fermat's Last Theorem? Gauss' friend H. W. M. Olbers wrote him a letter from Bremen on March 7, 1816, in which he told Gauss that the Paris Academy offered a big prize for anyone who would present a proof or a disproof of Fermat's Last Theorem. Gauss surely could use the money, his friend suggested. At that time, as he did throughout his mathematical career, Gauss received financial support from the Duke of Brunswick, which allowed him to do his mathematical work without the need for additional employment. But he was far from rich. And, as Olbers suggested, no other mathematician had nearly his expertise or ability. "It seems right to me, dear Gauss, that you should get busy about this," he concluded.

But Gauss would not be tempted. Possibly he knew how deceptive Fermat's Last Theorem really was. The great genius in number theory may have been the only mathematician in Europe to realize just how difficult it would be to prove. Two weeks later, Gauss wrote to Olbers his opinion of Fermat's Last Theorem: "I am very much obliged for your news concerning the Paris prize. But I confess that Fermat's Theorem as an isolated proposition has very little interest for me, because I could easily lay down a multitude of such propositions, which one could neither prove nor dispose of." Ironically, Gauss made great contributions to the branch of mathematics known as complex analysis—an area incorporating the imaginary numbers worked on by Euler. Imaginary numbers would have a decisive role in twentieth-century understanding of the context of Fermat's Last Theorem.

Imaginary Numbers

The complex number field is a field of numbers based on the usual real numbers, and what are called imaginary numbers, which were known to Euler. These numbers arose when mathematicians were looking to define as a number the solution of an equation such as $x^2 + 1 = 0$. There is no "real" solution to this simple equation, because there is no real number which, when squared, gives -1—the number that when added to 1 will give the answer zero. But if we could somehow define the square root of negative one as a number, then—while not a real number—it would be the solution to the equation.

The number line was therefore extended to include imaginary numbers. These numbers are multiples of the square root of -1, denoted by *i*. They were put on their own number line, perpendicular to the real number line. Together, these two axes give us the complex plane. The complex plane is shown below. It has many surprising properties, such as rotation being multiplication by *i*.

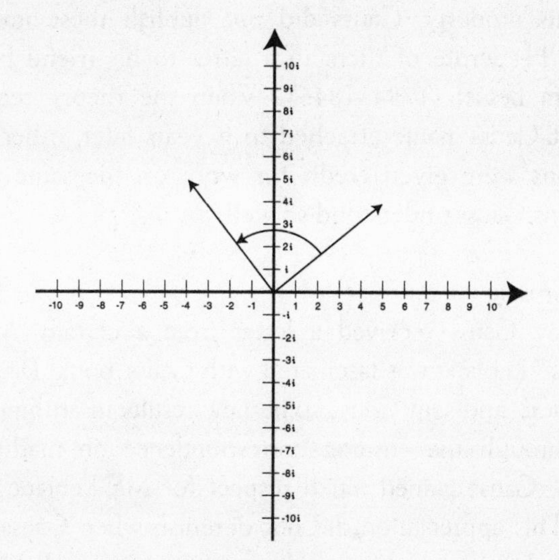

Multiplication by *i* rotates counterclockwise.

The complex plane is the smallest field of numbers that contains the solutions of all quadratic equations. It was found to be very useful, even in applications in engineering, fluid mechanics, and other areas. In 1811, decades ahead of his time, Gauss was studying the behavior of functions on the complex plane. He discovered some amazing properties of these functions, known as analytic functions. Gauss found that analytic functions had special smoothness, and they allowed for particularly neat calculations. Analytic functions preserved angles between lines and arcs on the plane—an aspect that would become significant in the twentieth century. Some analytic functions, called modular forms, would prove crucial in new approaches to the Fermat problem.

In his modesty, Gauss did not publish these impressive results. He wrote of them in a letter to his friend Friedrich Wilhelm Bessel (1784–1846). When the theory reemerged without Gauss' name attached to it years later, other mathematicians were given credit for work on the same analytic functions Gauss understood so well.

Sophie Germain

One day Gauss received a letter from a certain "Monsieur Leblanc." Leblanc was fascinated with Gauss' book, *Disquisitiones Arithmeticae*, and sent Gauss some new results in arithmetic theory. Through the ensuing correspondence on mathematical matters, Gauss gained much respect for Mr. Leblanc and his work. This appreciation did not diminish when Gauss discovered that neither was his correspondent's real name Leblanc, nor was the writer of the letters a "Mr." The mathematician writing so eloquently to Gauss was one of very few women active in the profession at that time, Sophie Germain (1776–1831). In fact, upon discovering the deception, Gauss wrote her:

> But how to describe to you my admiration and astonishment at seeing my esteemed correspondent Mr. Leblanc metamorphose himself into this illustrious personage who gives such a brilliant example of what I would find difficult to believe...

(A letter from Gauss to Sophie Germain, written from Brunswick on Gauss' birthday, as stated in French at the end of his letter: "Bronsvic ce 30 avril 1807 jour de ma naissance.")

Sophie Germain assumed a man's name to avoid the prejudice against women scientists prevalent in those days and to gain Gauss' serious attention. She was one of the most important mathematicians to attempt a proof of Fermat's Last Theorem, and to make considerable headway on the problem. Sophie Germain's Theorem, which gained her much recognition, states that if a solution of Fermat's equation with $n=5$ existed, then all three numbers had to be divisible by 5. The theorem separated Fermat's Last Theorem into two cases: Case I for numbers not divisible by 5, and Case II for numbers that are. The theorem was generalized to other powers, and Sophie Germain gave a general theorem which allowed for a proof of Fermat's Last Theorem for all prime numbers, n, less than 100 in Case I. This was an important result, which reduced the possible cases where Fermat's Last Theorem might fail for primes less than 100 to only Case II.[10]

Sophie Germain had to drop the disguise she was using when Gauss asked his friend "Leblanc" for a favor. In 1807, Napoleon was occupying Germany. The French were imposing war fines on the Germans, and were determining the amounts owed by each resident based on what they perceived each person was worth. As a prominent professor and astronomer at Göttingen, Gauss was determined to owe 2,000 francs—far beyond his means. A number of French mathematicians who were friends of the great Gauss offered their help, but he refused to take their money. Gauss wanted someone to intercede on his behalf with the French General Pernety in Hanover.

He wrote his friend, Monsieur Leblanc, to ask if Leblanc might contact the French General on Gauss' behalf. When Sophie Germain gladly complied, it became clear who she was. But Gauss was thrilled, as seen from his letter, and their correspondence continued and developed further on many mathematical topics. Unfortunately, the two never met. Sophie Germain died in Paris in 1831, before the University of Göttingen could award her the honorary doctorate which Gauss had recommended she receive.

Sophie Germain had many other achievements to her credit in addition to her contributions to the solution of Fermat's Last Theorem. She was active in the mathematical theories of acoustics and elasticity, and other areas in applied and pure mathematics. In number theory, she also proved theorems on which prime numbers can lead to solvable equations.

The Blazing Comet of 1811

Gauss did much important work in astronomy, determining the orbits of planets. On August 22, 1811, he first observed a comet barely visible in the night sky. He was able to predict the comet's precise trajectory toward the sun. When the comet became clearly visible and blazed across the sky, the superstitious and oppressed peoples of Europe saw in it a sign from heaven signifying the coming demise of Napoleon. Gauss saw in the comet a realization of the orbit he had predicted for it to exact numerical accuracy. But the unscientific masses were also right—the next year Napoleon was defeated and retreated from Russia. Gauss was amused. He was not unhappy to see the Emperor defeated after French forces had extorted such high amounts from him and his countrymen.

The Disciple

The Norwegian mathematician Niels Henrik Abel visited Paris in October of 1826. There, he tried to meet other mathematicians—Paris was at that time a mecca for mathematics. One of the people who impressed Abel the most was Peter Gustav Lejeune Dirichlet (1805–1859), a Prussian who was also visiting Paris and gravitated toward the young Norwegian, at first thinking he was a fellow Prussian. Abel was much impressed by the fact that Dirichlet had proved Fermat's Last Theorem for $n=5$. He wrote about it in a letter to a friend, mentioning that this was also done by Adrien-Marie Legendre (1752–1833). Abel described Legendre as extremely polite but very old. Legendre had proved the Fermat result for $n = 5$ independently of Dirichlet, two years after him. Unfortunately, this was always happening to Legendre—much of his work was superseded by that of younger mathematicians.

Dirichlet was a friend and disciple of Gauss. When Gauss' great book, the *Disquisitiones Arithmeticae*, was published, it quickly went out of print. Even mathematicians whose work was related to that of Gauss could not obtain a copy. And many who did, did not understand the depth of Gauss' work. Dirichlet had his own copy. He carried this copy of Gauss' book with him on his many travels, to Paris, Rome, and elsewhere on the Continent. Everywhere he went, Dirichlet slept with the book under his pillow. Gauss' book became known as the book of the seven seals: The gifted Dirichlet is known as the person who broke the seven seals. Dirichlet did more than anyone else to explain and interpret the book of his great master to the rest of the world.

Besides amplifying and explaining the *Disquisitiones*, as well as proving Fermat's Last Theorem for the power five, Dirichlet did other great mathematics. One interesting result that Dirichlet proved was connected with the following progression of numbers: a, $a + b$, $a + 2b$, $a + 3b$, $a + 4b$,... and so on, where the numbers a and b are integers that have no common divisor other than 1 (that is, they are numbers such as 2 and 3, or 3 and 5; rather than numbers such as 2 and 4, which have the common divisor 2, or the numbers 6 and 9, which have the common divisor 3). Dirichlet proved that this progression of numbers contains infinitely many prime numbers. The amazing element of Dirichlet's proof was that he had crafted it using an area of mathematics which seemed in those days to be very far from number theory, where this problem rightfully belongs. In his proof, Dirichlet used the field called analysis, an important area of mathematics which contains the calculus. Analysis deals with continuous things: functions on a continuum of numbers on the line, which seems very far from the discrete world of integers and prime numbers—the realm of number theory.

It would be a similar bridge across seemingly different branches of mathematics that would usher in the modern philosophy solving the Fermat mystery in our own century. Dirichlet was a bold pioneer in this area of unifying disparate branches of mathematics. The student later inherited the master's position. When Gauss died in 1855, Dirichlet left his prestigious job in Berlin to accept the honor of replacing Gauss at Göttingen.

Napoleon's Mathematicians

The Emperor of the French loved mathematicians, even if he wasn't one himself. Two who were especially close to him were Gaspard Monge (1746–1818), and Joseph Fourier (1768–1830). In 1798, Napoleon took the two mathematicians with him to Egypt, to help him "civilize" that ancient country.

Fourier was born in Auxerre, France, on March 21, 1768, but at the age of eight he was orphaned and was helped by the local Bishop to enter the military school. Even at the age of twelve, Fourier showed great promise and was writing sermons for church dignitaries in Paris, which they passed off as their own. The French Revolution of 1789 saved the young Fourier from life as a priest. Instead he became a professor of mathematics, and an enthusiastic supporter of the revolution. When the revolution gave way to the Terror, Fourier was repulsed by its brutality. He used his eloquence, developed over years of writing sermons for others, to preach against the excesses. Fourier also used his great public speaking skills in teaching mathematics at the best schools in Paris.

Fourier was interested in engineering, applied mathematics, and physics. At the École Polytechnique, he did serious research in these areas, and many of his papers were presented to the Academy. His reputation brought him to the attention of Napoleon, and in 1798 the Emperor asked Fourier to accompany him, aboard his flagship, along with the French fleet of five hundred ships headed for Egypt. Fourier was to be part of the Legion of Culture. The Legion's charge was to "endow the people of Egypt with all the benefits of European civilization." Culture was to be brought to these people while they were being conquered by the invading armada. In Egypt, the two mathematicians founded the Egyptian Institute, and Fourier stayed there until 1802, when he returned to France and was made a prefect of the region around Grenoble. There he was responsible for many good public works such as draining marshlands and stamping out malaria. With all that work, Fourier, the mathematician-turned-administrator, managed to find time to do his best mathematical thinking. Fourier's masterpiece was the mathematical theory of heat, answering the important question: How is heat conducted? This work won him a Grand Prize from the Academy in 1812. Part of his work was based on experiments he had done in the deserts of Egypt during his years there. Some of his friends believed that these experiments, including putting himself through the intense heat that developed in closed rooms, contributed to his early death at the age of 62.

Fourier spent the last years of his life telling tales about Napoleon and his close association with him, both in Egypt and after Napoleon's escape from Elba. Fourier's research on heat, however, was what immortalized him, for he developed an important theory of periodic functions. A series of such periodic functions, when used in a particular way to estimate another function, is called a Fourier Series.

Periodic Functions

The clearest example of a periodic function is your watch. Minute by minute, the large hand moves around a circle, and after sixty minutes it comes back to the exact same place where it started. Then it continues and in exactly sixty minutes it again returns to the same spot. (Of course the little hand will have changed positions as the hours go by.) The minutes hand on a watch is a periodic function. Its period is exactly sixty minutes. In a sense, the space of all minutes of eternity—the set of infinitely many minutes from now until forever—can be *wrapped* by the big hand of a watch on the outer edge of the face of the watch:

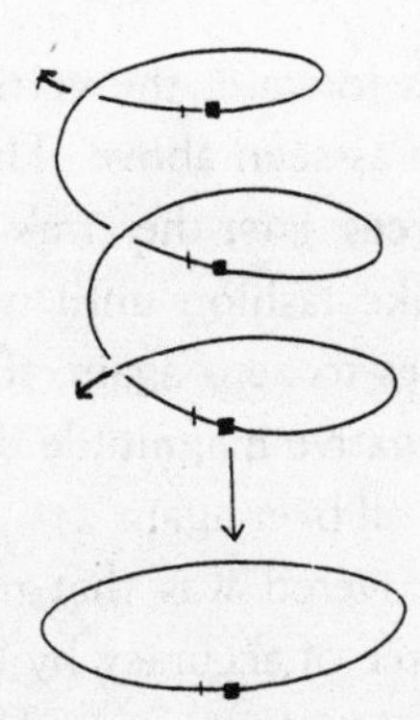

To take another example, on a locomotive speeding on the railroad track the arm transferring power from the engine to the wheel moves up and down along that wheel as it rotates. Every full circle of the wheel, the arm comes back to its original position—this arm, too, is periodical. The vertical height of the arm, when the radius of the train wheel is one unit, is defined as the sine function. This is the elementary trigonometric function taught at school. The cosine function is the horizontal measure of the arm. Both sine and cosine are functions of the angle the arm makes with a horizontal line through the center of the wheel. This is shown below.

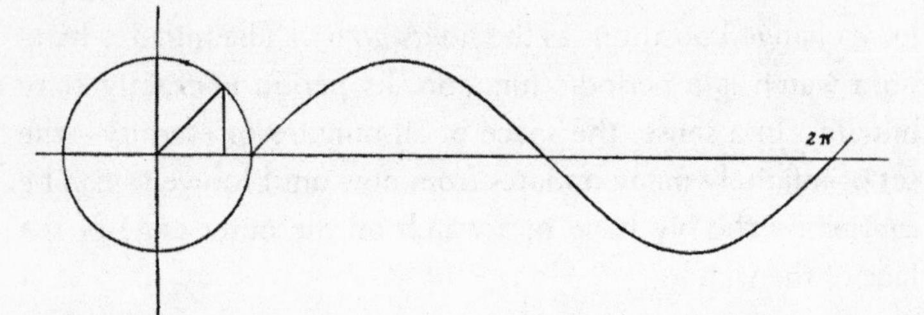

As the train moves forward, the vertical height of the arm traces a wave-pattern as seen above. This pattern is periodic. Its period is 360 degrees. First the arm's height is zero, then it goes up in a wave-like fashion until it reaches one, then it decreases, then it goes to zero again, then to negative values until one, then its negative magnitude decreases to zero. And then the cycles starts all over again.

What Fourier discovered was that most functions can be estimated to any degree of accuracy by the sum of many (theoretically infinitely many, for near-perfect accuracy) sine and cosine functions. This is the famous result of the Fourier series. This expansion of any function in terms of the sum of many sine and cosine functions is useful in many applications of mathematics when the actual mathematical expression of interest is difficult to study but the sum of the sines and cosines, all multiplied by different factors, can be easily manipulated and evaluated—and this is especially practical on a computer. The area of mathematics known as numerical analysis is concerned with computer methods for evaluating functions and other numerical quantities. Fourier analysis is a substantial part of numerical analysis, and consists of techniques for studying difficult problems, many with no closed-form (that is, given by a simple mathematical expression) solution using the Fourier series of periodic function. After Fourier's pioneering work, expansions using other simple functions, mostly polynomials (that is, increasing powers of a variable: squares, cubes, and so on) were also developed. When your calculator computes the square root of a number, it is doing so by an approximation based on such a method. The Fourier series of sines and cosines is especially useful in estimating phenomena that are naturally the sum of periodic elements—for example, music. A musical piece may be decomposed into its harmonics. The tides, the phases of the moon, and sun spots are examples of simple periodic phenomena.

While the applications of Fourier's periodic functions to natural effects and computational methods are of great importance, the surprising fact is that Fourier series and analysis found some useful applications in pure mathematics, a field that was never one of Fourier's main interests. In the twentieth century, Fourier series found a role in number theory as the tool for transforming mathematical elements from one area to another in the work of Goro Shimura. (The proof of Shimura's conjecture was at the crux of proving Fermat's Last Theorem.) Extensions of the periodic functions of Fourier to the complex plane, linking together these two areas of mathematics, would lead to the discovery of automorphic functions and modular forms—which also had a crucial impact on Fermat's Last Theorem through the early twentieth century work of another French mathematician, Henri Poincaré.

A Lame Proof

At the March 1, 1847 meeting of the Paris Academy, the mathematician Gabriel Lamé (1795–1870) announced very excitedly that he had obtained a general proof of Fermat's Last Theorem. Until then, only specific powers, n, were attacked, and a proof had been given of the theorem in the cases $n=3$, 4, 5, 7. Lamé suggested that he had a general approach to the problem, which would work for any power n. Lamé's method was to factor the left side of Fermat's equation, $x^n + y^n$, into linear factors using complex numbers. Lamé then went on to state modestly that the glory should not all go to him, since the method he suggested was introduced to him by Joseph Liouville (1809–1882). But Liouville took the podium after Lamé and brushed aside any praise. Lamé had not proved Fermat's Last Theorem, he quietly stated, because the factorization he suggested was not unique (that is, there were many ways to carry out the factorization, so there was no solution). It was a gallant attempt, one of many, but it did not bear fruit. However, the idea of factorization, that is, breaking down the equation to a product of factors, was to be tried again.

Ideal Numbers

The person to again try factorization was Ernst Eduard Kummer (1810–1893)—the man who got closer than anyone else in his time to a general solution of Fermat's problem. Kummer, in fact, invented an entire theory in mathematics, the theory of ideal numbers, in attempting to prove Fermat's Last Theorem.

Kummer's mother, a widow when her son was only three, worked hard to assure her boy a good education. At the age of eighteen he entered the University of Halle, in Germany, to study theology and prepare himself for life in the church. An insightful professor of mathematics with enthusiasm for algebra and number theory got the young Kummer interested in these areas, and he soon abandoned theology for mathematics. In his third year as a student, the young Kummer solved a difficult problem in mathematics for which a prize had been offered. Following this success, he was awarded a doctorate in mathematics at the age of twenty-one.

But Kummer was unable to find a university position and therefore had to take a teaching job at his old gymnasium (high school). He remained a schoolteacher for another ten years. Throughout this period, Kummer did much research, which he published and wrote in letters to several leading mathematicians of his time. His friends realized how sad it was to have such a gifted mathematician spend his life teaching high school math. With the help of some eminent mathematicians, Kummer was given the position of professor at the University of Breslau. A year later, in 1855, Gauss died. Dirichlet had taken Gauss' place at Göttingen, leaving behind his old job at the prestigious University of Berlin. Kummer was chosen to replace Dirichlet in Berlin. He remained in that position until his retirement.

Kummer worked on a wide range of problems in mathematics, ranging from very abstract to very applied—even applications of mathematics in warfare. But he made his name for his extensive work on Fermat's Last Theorem. The French mathematician Augustin-Louis Cauchy (1789–1857), thought on a number of occasions that he had found a general solution to the Fermat problem. But the restless, careless Cauchy realized in every attempt that the problem was much bigger than he had assumed. The fields of numbers he was working with always failed to have the property he required. Cauchy left the problem and went to work on other things.

Kummer became obsessed with Fermat's Last Theorem, and his attempts at a solution led him down the same futile track taken by Cauchy. But instead of giving up hope when he recognized that the number fields involved failed to have some property, he instead *invented* new numbers that would have the property he needed. These numbers he called "ideal numbers." Kummer thus developed an entirely new theory from scratch, which he used in his attempts to prove Fermat's Last Theorem. At one point, Kummer thought he finally had a general proof, but this unfortunately fell short of what was needed.

Nonetheless, Kummer did achieve tremendous gains in his attack on the Fermat problem. His work with ideal numbers enabled him to prove Fermat's Last Theorem for a very extensive class of prime numbers as the exponent n. Thus, he was able to prove that Fermat's Last Theorem was true for an infinite number of exponents, namely those that are divisible by "regular" primes. The "irregular" prime numbers eluded him. The only irregular primes less than 100 are 37, 59, and 67. Kummer then worked separately on the problem of these irregular prime numbers and was eventually able to prove Fermat's theorem for these numbers. By the 1850s, using Kummer's incredible breakthroughs, Fermat's Last Theorem was known to be true for all exponents less than $n = 100$, as well as infinitely many multiples of prime numbers in this range. This was quite an achievement, even if it was not a general proof and still left out infinitely many numbers for which it was not known whether the theorem held true.

In 1816, the French Academy of Sciences offered a prize to anyone who would prove Fermat's Last Theorem. In 1850 the Academy again offered a gold medal and 3,000 francs to the mathematician who would prove Fermat's Last Theorem. In 1856, the Academy decided to withdraw the award, since it didn't seem like a solution to Fermat's problem was imminent. Instead, the Academy decided to give the award to Ernst Eduard Kummer, "for his beautiful researches on the complex numbers composed of roots of unity and integers." Thus Kummer was awarded a prize for which he never even applied.

Kummer continued his tireless efforts on Fermat's Last Theorem, stopping his research only in 1874. Kummer also did pioneering work on the geometry of four-dimensional space. Some of his results are useful in the field of modern physics known as quantum mechanics. Kummer died of the flu in his eighties, in 1893.

Kummer's success with ideal numbers is even more highly praised by mathematicians than the actual advances he made in the solution of Fermat's problem using these numbers. The fact that this notable theory was inspired by attempts to solve Fermat's Last Theorem shows how new theories can be developed by attempts to solve particular problems. In fact, Kummer's theory of ideal numbers led to what are now known as "ideals," which have had an impact on the work of Wiles and other mathematicians on Fermat's theorem in the twentieth century.

Another Prize

In 1908, the Wolfskehl Prize of one hundred thousand marks was offered in Germany for anyone who would come up with a general proof of Fermat's Last Theorem. In the first year of the prize, 621 "solutions" were submitted. All of them were found to be false. In the following years, thousands more "solutions" were submitted, with the same effect. In the 1920s, German hyperinflation reduced the real value of the 100,000 marks to nothing. But false proofs of Fermat's Last Theorem continued to pour in.

Geometry Without Euclid

New developments started taking place in mathematics in the nineteenth century. Janos Bolyai (1802–1860) and Nicolas Ivanovitch Lobachevsky (1793–1856), a Hungarian and a Russian respectively, changed the face of geometry. By doing away with Euclid's axiom that two parallel lines never meet, these two were able independently to formulate a geometrical universe which maintained many of Euclid's properties but allowed two parallel lines to meet at a point at infinity. The new geometry which came about can be seen, for example, on a sphere such as the globe. Locally, two longitude lines are parallel. But in reality, as they are followed to the North Pole, the two lines meet there. The new geometry solves many problems and explained situations which, until that time, seemed mysterious and without a solution.

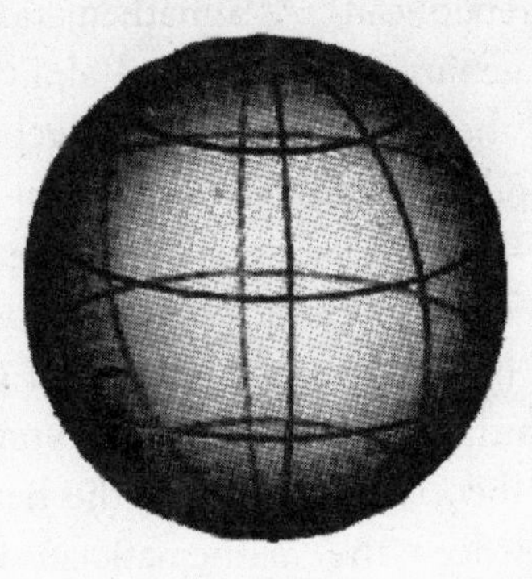

Beauty and Tragedy

Abstract algebra, a field derived from the familiar algebra taught in school as a system for solving equations, was developed in the nineteenth century. In this area, the beautiful theory of Galois stands out.

Évariste Galois was born in 1811 in the small village of Bourg-la-Reine, outside Paris. His father was the mayor of the town and a staunch republican. Young Évariste grew up on the ideals of democracy and freedom. Unfortunately, most of France was at that time heading in the opposite direction. The French Revolution had come and gone, and so had Napoleon. But the dreams of freedom, equality, and fraternity had not yet been achieved. And the royalists were enjoying a comeback in France, with a Bourbon once again crowned King of the French—now to rule together with the people's representatives.

Évariste's life was steeped in the lofty ideals of the revolution. He was a good ideologue, and he gave some moving speeches to the republicans. As a mathematician, on the other hand, he was a genius of unparalleled ability. As a teenager, Galois absorbed the entire theories of algebra and equations known to accomplished mathematicians of his day, and—while still a schoolboy—developed his own complete system, known today as Galois Theory. Unfortunately, he was not to enjoy any recognition in his tragically short life. Galois stayed up nights at his boarding school, while everyone else slept, and wrote down his theory. He sent it to the head of the French Academy of Science—the mathematician Cauchy—hoping that Cauchy would help him publish the theory. But Cauchy was not only very busy, he was also arrogant and careless. And Galois' brilliant manuscript ended up in a trash can, unread.

Galois tried again, with the same result. In the meantime, he failed to pass the entry exams to the École Polytechnique, where most of the celebrated mathematicians of France got their training. Galois had a habit of always working things out in his head. He never took notes or wrote things down until he had actual results. This method concentrated on *ideas* rather than detail. Young Evariste had little patience for, or interest in, the details. It was the great idea, the beauty of the larger theory, that interested him. Consequently, Galois was not at his best when taking an examination in front of a blackboard. And this is what caused him to fail twice in his attempts to gain entrance to the school of his dreams. Twice in front of the blackboard he did not perform well writing things down, and he got irritated when asked for details he just didn't consider important. It was a tragedy that an incredibly intelligent young person would be questioned by much less able examiners who could not understand his deep ideas and took his reluctance to give trivial details for ignorance. When he realized he was going to fail the second and last permissible attempt, and that the gates of the school would be closed to him forever, Galois threw the blackboard eraser in the face of his examiner.

Galois had to make do with a second-best choice, the École Normale. But even there he did not fare well. Galois' father, the mayor of Bourg-la-Reine, was the target of clerical intrigues in the town. An unscrupulous priest circulated pornographic verses, signing them with the mayor's name. Months of such persecution caused Galois' father to lose his confidence and he became convinced that the world was out to get him. Slowly losing touch with reality, he went to Paris. There, in an apartment not far from where his son was studying, the father killed himself. The young Galois never recovered from this tragedy. Obsessed with the lost cause of the 1830 revolution, and frustrated with the school director, whom he considered an apologist for the royalists and clerics, Galois wrote a scathing letter criticizing the director. He was inspired to do so after three days of rioting in the streets, when students all over Paris were revolting against the regime. Galois and his classmates were kept locked in their school, unable to scale the tall fence. The angry Galois sent his blistering letter criticizing the school director to the *Gazette des Écoles*. As a result, he was expelled from the school. But Galois was undaunted—he wrote a second letter to the *Gazette*, and addressed the students at the school to speak up for honor and conscience. He got no response.

Out of school, Galois started offering private lessons in mathematics. He wanted to teach his own mathematical theories, outside the French schools, when he was all of nineteen years old. But Galois could not find students to teach—his theories were too advanced; he was far ahead of his time.

Facing an uncertain future and doomed not to be able to pursue a decent education, in desperation Galois joined the artillery branch of the French National Guard. Among the National Guard, headed in its past by Lafayette, there were many liberal elements close to young Galois' political philosophy. While in the Guard, Galois made one last attempt to publish his mathematical work. He wrote a paper on the general solution of equations—today recognized as the beautiful Galois Theory—and sent it to Siméon-Denis Poisson (1781–1840) at the French Academy of Sciences. Poisson read the paper but determined that it was "incomprehensible." Once again, the nineteen-year-old was so far ahead of any of the older French mathematicians of his day that his elegant new theories went way over their heads. At that moment, Galois decided to abandon mathematics and to become a full-time revolutionary. He said that if a body was needed to get the people involved with the revolution, he would donate his.

On May 9, 1831, two hundred young republicans held a banquet in which they protested against the royal order disbanding the artillery of the National Guard. Toasts were offered to the French Revolution and its heroes, as well as to the new revolution of 1830. Galois stood up and proposed a toast. He said "To Louis Philippe," the Duke of Orléans, who was now King of the French. While saying this, holding up his glass, Galois was also holding up an open pocket-knife in his other hand. This was interpreted as a threat on the life of the King, and caused a riot. The next day Galois was arrested.

In his trial for threatening the life of the King, Galois' attorney claimed that Galois had actually said, while holding his knife, "To Louis Philippe, if he becomes a traitor." Some of Galois' artillery friends who were present testified to this, and the jury found him not guilty. Galois retrieved his knife from the evidence table, closed it and put it in his pocket, and left the courtroom a free man. But he was not free for very long. A month later he was arrested as a dangerous republican, and kept in jail without a charge while the authorities looked for a charge against him that would stick. They finally found one—wearing the uniform of the disbanded artillery. Galois was tried for this charge and sentenced to six months in jail. The royalists were pleased to finally put away a twenty-year-old they considered to be a dangerous enemy of the regime. Galois was paroled after some time and was moved to a halfway facility. What happened next is open to question. While on parole, Galois met a young woman and fell in love. Some believe he was set up by his royalist enemies who wanted to put an end to his revolutionary activities once and for all. At any rate, the woman with whom he got involved was of questionable virtue (*"une coquette de bas étage"*). As soon as the two became lovers, a royalist came to "save her honor" and challenged Galois to a duel. The young mathematician was left no way out of the mess. He tried everything he could to talk the man out of the duel, but to no avail.

The night before the duel, Galois wrote several letters.

These letters to his friends lend support to the theory that he was framed by the royalists. He wrote that he was challenged by two of the royalists and that they had put him on his honor not to tell his republican friends about the duel. "I die the victim of an infamous coquette. It is in a miserable brawl that my life is extinguished. Oh, why die for so trivial a thing, why die for something so despicable!" But most of that last night before the duel, Galois carefully put down on paper his entire mathematical theory, and sent it to his friend Auguste Chevalier. At dawn on May 30, 1832, Galois faced his challenger on a deserted field. He was shot in the stomach and left lying in agony alone in the field. No one bothered to call a doctor. Finally a peasant found him and brought him to the hospital, where he died the next morning. He was twenty years old. In 1846, the mathematician Joseph Liouville edited and published Galois' elegant mathematical theory in a journal. Galois' theory would supply the crucial step in the method used a century and a half later in attacking Fermat's Last Theorem.

Another Victim

Cauchy's carelessness and arrogance ruined the life of at least one other brilliant mathematician. Niels Henrik Abel (1802–1829) was the son of the pastor of the village of Findö in Norway. When he was sixteen, a teacher encouraged Abel to read Gauss' famous book, the *Disquisitiones*. Abel was even successful in filling in some gaps in the proofs of some of the theorems. But two years later, his father died and young Abel had to postpone his study of mathematics and concentrate his efforts on supporting the family. In spite of the great difficulties he faced, Abel managed to continue some study of mathematics and made a remarkable mathematical discovery at the age of nineteen. He published a paper in 1824 in which he proved that no solution was possible for equations of the fifth degree. Abel had thus solved one of the most celebrated problems of his day. Yet the gifted young mathematician had still not been offered an academic position, which he badly needed to support his family, and so he sent his work to Cauchy for evaluation and possible publication and recognition. The paper Abel sent Cauchy was of extraordinary power and generality. But Cauchy lost it. When it appeared in print, years later, it was much too late to help Abel. In 1829 Abel died from tuberculosis, brought on by poverty and the strain of supporting his family in dire circumstances. Two days after his death, a letter arrived addressed to Abel informing him he had been appointed professor at the University of Berlin.

The concept of an Abelian Group (now considered a word and written with a small "a," abelian) is very important in modern algebra and is a crucial element in the modern treatment of the Fermat problem. An abelian group is one where the order of mathematical operations can be reversed without affecting the outcome. An abelian variety is an even more abstract algebraic entity, and its use was also important in modern approaches to the solution of Fermat's Last Theorem.

Dedekind's Ideals

The legacy of Carl Friedrich Gauss continued through the centuries. One of Gauss' most notable mathematical successors was Richard Dedekind (1831–1916), born in the same town as the great master, in Brunswick, Germany. Unlike Gauss, however, as a child Dedekind showed no great interest in or capacity for mathematics. He was more interested in physics and chemistry, and saw mathematics as a servant to the sciences. But at the age of seventeen, Dedekind entered the same school where the great Gauss got his mathematical training—Caroline College—and there his future changed. Dedekind became interested in mathematics and he pursued that interest in Göttingen, where Gauss was teaching. In 1852, at the age of 21, Richard Dedekind received his doctorate from Gauss. The master found his pupil's dissertation on the calculus "completely satisfying." This was not such a great compliment, and in fact Dedekind's genius had not yet begun to manifest itself.

In 1854, Dedekind was appointed a lecturer in Göttingen. When Gauss died in 1855 and Dirichlet moved from Berlin to take his position, Dedekind attended all of Dirichlet's lectures at Göttingen and edited the latter's pioneering treatise on number theory, adding a supplement based on his own work. This supplement contained an outline of the theory Dedekind developed for algebraic numbers, which are defined as solutions of algebraic equations. They contain multiples of square roots of numbers along with rational numbers. Algebraic number fields are very important in the study of Fermat's equation, as they arise from the solution of various kinds of equations. Dedekind thus developed a significant area within number theory.

Dedekind's greatest contribution to the modern approach to Fermat's last Theorem was his development of the theory of ideals, abstractions of Kummer's ideal numbers. A century after their development by Dedekind, ideals would inspire Barry Mazur, and Mazur's own work would be exploited by Andrew Wiles.

In the 1857–8 academic year, Richard Dedekind gave the first mathematics course on Galois theory. Dedekind's understanding of mathematics was very abstract, and he elevated the theory of groups to the modern level at which it is understood and taught today. Abstraction made the twentieth-century approach to Fermat's problem possible. Dedekind's groundbreaking course on the theories developed by Galois was a great step in this direction. The course was attended by two students.

Then Dedekind's career took a strange turn. He left Göttingen for a position in Zürich, and after five years, in 1862, he returned to Brunswick, where he taught at a high school for fifty years. No one has been able to explain why a brilliant mathematician who brought algebra to an incredibly high level of abstraction and generality suddenly left one of the most prestigious positions at any European university to teach at an unknown high school. Dedekind never married and lived for many years with his sister. He died in 1916, and maintained a sharp, active mind to his last day.

Fin de Siècle

At the turn of the nineteenth century, there lived in France a mathematician of great ability in a surprisingly wide variety of areas. The breadth of knowledge of Henri Poincaré (1854–1912) extended beyond mathematics. In 1902 and later, when he was already a renowned mathematician, Poincaré wrote popular books on mathematics. These paperbacks, read by people of all ages, were a common sight in the cafés and the parks of Paris.

Poincaré was born to a family of great achievers. His cousin, Raymond Poincaré, rose to be the president of France during the First World War. Other family members held government and public service positions in France as well.

From a young age, Henri displayed a powerful memory. He could recite from any page of a book he read. His absentmindedness, however, was legendary. A Finnish mathematician once came all the way to Paris to meet with Poincaré and discuss some mathematical problems. The visitor was kept waiting for three hours outside Poincaré's study while the absentminded mathematician paced back and forth—as was his habit throughout his working life. Finally, Poincaré popped his head into the waiting room and exclaimed: "Sir, you are disturbing me!" upon which the visitor summarily left, never to be seen in Paris again.

Poincaré's brilliance was recognized when he was in elementary school. But since he was such a universalist—a renaissance man in the making—his special aptitude for mathematics did not yet manifest itself. He distinguished himself with his excellence in writing at an early age. A teacher who recognized and encouraged his ability treasured his school papers. At some point, however, the concerned teacher had to caution the young genius: "Don't do so well, please...try to be more ordinary." The teacher had good reason to make this suggestion. Apparently, French educators had learned something from the misfortunes of Galois half a century earlier—teachers found that gifted students often failed at the hands of uninspired examiners. His teacher was genuinely worried that

Poincaré was so brilliant that he might fail those exams. As a child, Poincaré was already absentminded. He often skipped meals because he forgot whether or not he had eaten.

Young Poincaré was interested in the classics and he learned to write well. As a teenager he became interested in mathematics and immediately excelled in it. He would work out problems entirely in his head while pacing in a room—then sit down and write everything very hastily. In this he resembled Galois and Euler. When Poincaré finally took his exams, he almost failed in math, as his elementary school teacher had feared years earlier. But he did pass, only because—at seventeen—his renown as a mathematician was so great that the examiners didn't dare fail him. "Any student other than Poincaré would have been given a failing grade," the chief examiner declared as he passed the student who went on to the École Polytechnique and became the greatest French mathematician of his time.

Poincaré wrote scores of books on mathematics, mathematical physics, astronomy, and popular science. He wrote research papers of over five hundred pages on new mathematical topics he developed. He made major contributions to topology, the area started by Euler. However, Poincaré's results were so important that this branch of mathematics is considered to have truly been launched only in 1895, with the publication of Poincaré's *Analysis Situs*. Topology—the study of shapes and surfaces and continuous functions—was important in understanding Fermat's problem in the late twentieth century. But even more essential to the modern approach to Fermat's Last Theorem was another area pioneered by Henri Poincaré.

Modular Forms

Poincaré studied periodic functions, such as the sines and cosines of Fourier—not on the number line as Fourier had done, but in the complex plane. The sine function, *sin x*, is the vertical height on a circle with radius 1 when the angle is *x*. This function is periodic: it repeats itself over and over every time the angle completes a multiple of its period, 360 degrees. This periodicity is a symmetry. Poincaré examined the complex plane, which contains real numbers on the horizontal axis and imaginary ones on the vertical one as shown below.

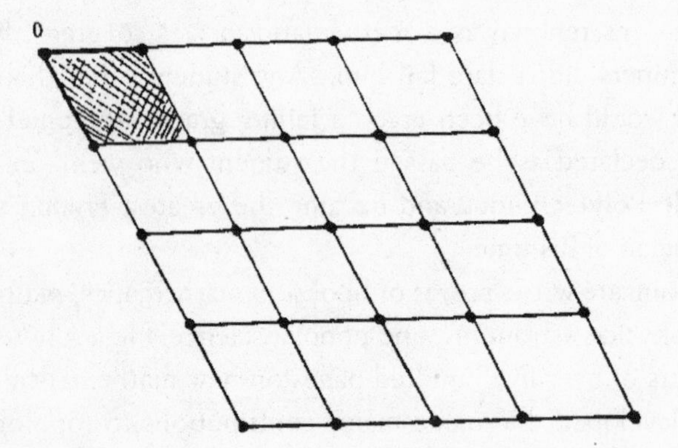

Here, a periodic function could be conceived as having a periodicity both along the real axis *and* along the imaginary axis. Poincaré went even further and posited the existence of functions with a wider array of symmetries. These were functions that remained unchanged when the complex variable *z* was changed according to $f(z)$———>$f(az+b/cz+d)$. Here the elements *a, b, c, d*, arranged as a matrix, formed an algebraic *group*. This means that there are infinitely many possible variations. They all commute with each other and the function f is *invariant* under this group of transformations. Poincaré called such weird functions automorphic forms.

The automorphic forms were very, very strange creatures since they satisfied many internal symmetries. Poincaré wasn't quite sure they existed. Actually, Poincaré described his research as follows. He said that for fifteen days he would wake up in the morning and sit at his desk for a couple of hours trying to convince himself that the automorphic forms he invented couldn't possibly exist. On the fifteenth day, he realized that he was wrong. These strange functions, hard to imagine visually, did exist. Poincaré extended them to even more complicated functions, called modular forms. The modular forms live on the upper half of the complex plane, and they have a hyperbolic geometry. That is, they live in a strange space where the non-Euclidean geometry of Bolyai and Lobachevsky rules. Through any point in this half-plane, many "lines" parallel to any given line exist.

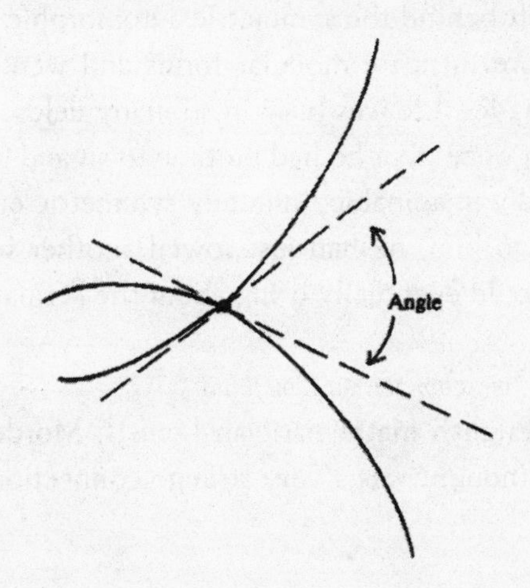

The very strange modular forms are symmetric in many ways within this space. The symmetries are obtained by adding a number to the function, and inverting it as 1/z. The tiling of the complex half-plane using these symmetries is shown below.

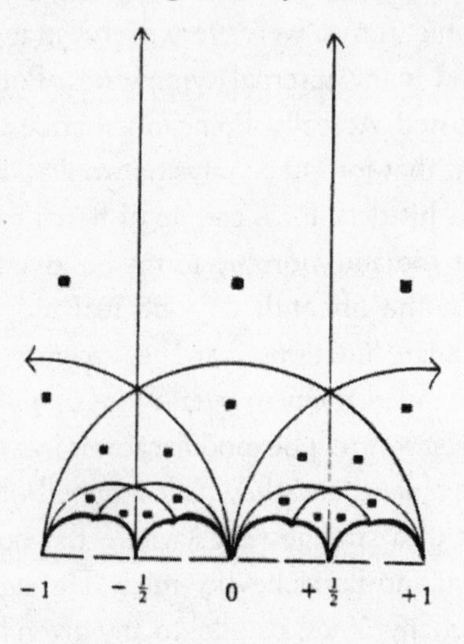

Poincaré left behind the symmetric automorphic forms and their even more intricate modular forms and went on to do other mathematics. He was busy in so many fields, often in a few of them at once, that he had no time to sit and ponder the beauty of hardly imaginable, infinitely symmetric entities. But unbeknownst to him, he had just sowed another seed in the garden that would eventually bring about the Fermat solution.

An Unexpected Connection with a Doughnut

In 1922, the English mathematician Louis J. Mordell discovered what he thought was a very strange connection between the solutions of algebraic equations and topology. The elements of topology are surfaces and spaces. These surfaces could be in any dimension: two dimensions, such as the figures in ancient Greek geometry, or they could be in three-dimensional space, or more. Topology is a study of continuous functions that act on these spaces, and the properties of the spaces themselves. The part of topology which concerned Mordell was the one of surfaces in three-dimensional space. A simple example of such a surface is a sphere: it is the surface of a ball, such as a basketball. The ball is three-dimensional, but its surface (assuming no depth) is a two-dimensional object. The surface of the earth is another example. The earth itself is three dimensional: any place on the earth or inside it can be given by its longitude (one dimension), its latitude (a second dimension), and its depth (the third dimension). But the surface of the earth (no depth) is two-dimensional, since any point on the surface of the earth can be specified by two numbers: its longitude and its latitude.

Two-dimensional surfaces in three-dimensional space can be classified according to their genus. The genus is the number of holes in the surface. The genus of a sphere is zero since there are no holes in it. A doughnut has one hole in it. Therefore the genus of a doughnut (mathematically called a torus) is one. A hole means a hole that runs completely through the surface. A cup with two handles has two holes through it. Therefore it is a surface of genus two.

A surface of one genus can be transformed by a continuous function into another surface of the same genus. The only way to transform a surface of one genus to one of a different genus

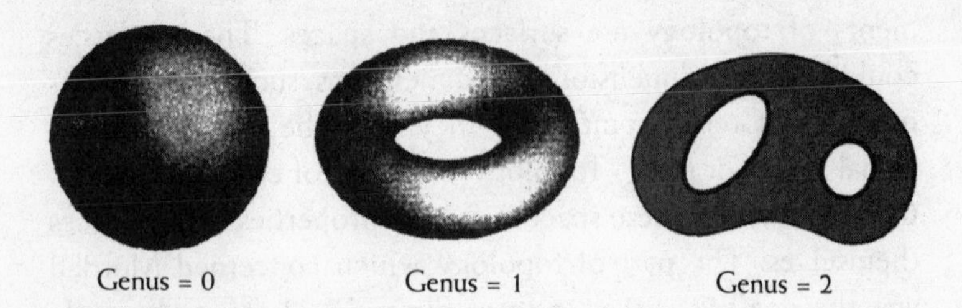

is by closing or opening some holes. This cannot be done by a continuous function, since it will require some ripping or fusing together, each of which is a mathematical discontinuity.

Mordell discovered a strange and totally unexpected connection between the number of holes in the surface (the genus) of the space of solutions of an equation and whether or not the

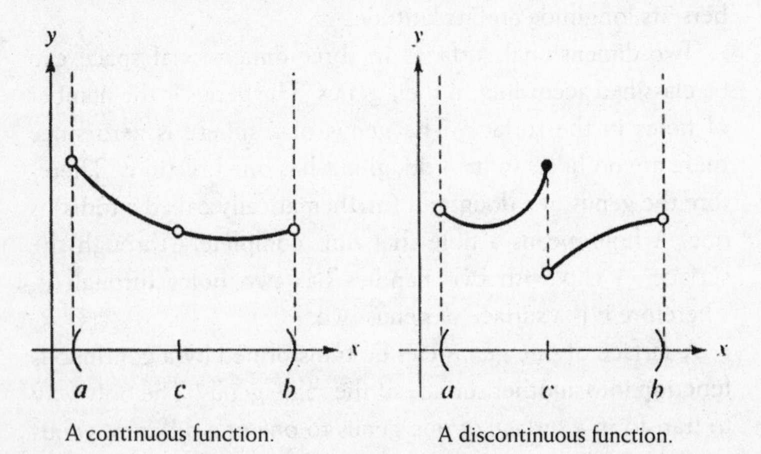

A continuous function. A discontinuous function.

equation had a finite number or an infinite number of solutions. If the surface of solutions, using complex numbers for greatest generality, had two or more holes (that is, had genus two or higher), then the equation had only finitely many whole-number solutions. Mordell was unable to prove his discovery, and it became known as Mordell's conjecture.

Faltings' Proof

In 1983, a twenty-seven year old German mathematician, Gerd Faltings, at that time at the University of Wuppertal, was able to prove the Mordell conjecture. Faltings was not interested in Fermat's Last Theorem, considering it an isolated problem in number theory. Yet his proof, which used great ingenuity along with the powerful machinery of algebraic geometry developed in this century, had profound implications for the status of Fermat's Last Theorem. Because the genus of the Fermat equation for n greater than 3 was 2 or more, it became clear that if integer solutions to the Fermat equation existed at all, then these were finite (which was comforting, since their number was now limited). Soon afterwards, two mathematicians, Granville and Heath-Brown, used Faltings' result to show that the number of solutions of Fermat's equation, if they existed, decreased as the exponent n increased. It was shown that the proportion of exponents for which Fermat's Last Theorem was true approached one hundred percent as n increased.

In other words, Fermat's Last Theorem was "almost always" true. If solutions to Fermat's equation existed (in which case the Theorem was not true), then such solutions were few and very far between. So the status of Fermat's Last Theorem in 1983 was the following. The theorem was proven for *n* up to a million (and in 1992 the limit was raised to 4 million). In addition, for larger *n*, if solutions existed at all then they were very few and decreasing with *n*.

The Mysterious Greek General with the Funny Name

There are dozens of excellent books on mathematics published in France, written in French, with the author listed as Nicolas Bourbaki. There was a Greek general named Bourbaki (1816–1897). In 1862 Bourbaki was offered the throne of Greece, which he declined. The general had an important role in the Franco-Prussian war, and there is a statue of him in the French city of Nancy. But General Bourbaki knew nothing about mathematics. And he never wrote a book, about mathematics or anything else. Who wrote the many volumes of mathematics bearing his name?

The answer lies in the happy days in Paris of between the two World Wars. Hemingway, Picasso and Matisse were not the only people who liked to sit in cafés and meet their friends and see people and be seen. At that time, around the same cafés on the Left Bank by the University of Paris, there flourished a vibrant mathematical community. Professors of mathematics from the university also liked to meet their friends, drink a café au lait or a pastis in a good brasserie on the Boulevard St. Michel by the beautiful Luxembourg Gardens, and discuss . . . mathematics. Springtime in Paris inspired writers, artists, and mathematicians. One imagines that on a sunny day at a pleasant café a rowdy group of mathematicians congregated. Feelings of fraternity overcame them as they argued animatedly over some fine points of a theory. Their revelry probably disturbed Hemingway, who writes he always liked to work by himself at a café, and he probably left to go to one of his alternate, less favorite haunts. But the mathematicians didn't care. They valued each other's company, and a café full of mathematicians—all speaking the same language of numbers and symbols and spaces and functions—was intoxicating. "This is what the Pythagoreans must have felt when they talked mathematics," perhaps one of the senior people in the group said, as he lifted his glass in a toast. "Yes, but they didn't drink Pernod," said another and everyone laughed. "But we could be like them," answered the first. "Why don't we form our *own* society? A secret one, naturally." There were voices of approval all around. Someone suggested they borrow the name of old General Bourbaki. There was a reason for this suggestion. In those days, the mathematics department at the University of Paris had an annual tradition of inviting a professional actor who presented himself to the assembled faculty and graduate students as Nicolas Bourbaki. He would then do a one-man show consisting of a long monologue of mathematical double-talk. Such presentations were very entertaining since the richness of modern mathematical theory makes for a vast vocabulary that is both descriptive of mathematics and has different meanings in everyday life. One such word is "dense." The rational numbers are said to be dense within the real numbers. This means that within any neighborhood of rational numbers are irrational numbers. But "dense" means many other things in everyday life.

Graduate students today like to play the same games of double meaning, and they like to tell the story of beautiful Poly Nomial who meets the smooth operator Curly Pi (polynomial, smooth operator, and curly pi are all mathematical terms).

And so the books these mathematicians wrote together bore the name Nicolas Bourbaki. A Bourbaki Seminar was initiated, where mathematical ideas and theories were discussed not infrequently. Membership in the society was supposed to be anonymous, and mathematical results were to be credited to the society in the name of Bourbaki, rather than to individual members.

But the members of Bourbaki were not the Pythagoreans. While the author of the textbooks was Nicolas Bourbaki, research results such as theorems and their proofs—which are far more prestigious than books—were credited to the individual mathematicians who achieved the results. One of the first members of Bourbaki was André Weil (1906–), who later moved to the United States and the Institute of Advanced Study at Princeton. His name would never be too far away from the important conjecture leading to the solution of the Fermat problem.

Another of the founders of Bourbaki was the French mathematician Jean Dieudonné, who like most of the other "French only" members of the society moved on to greener pastures at the universities in the United States. Dieudonné, who was the principal author of many of the books bearing the collective name of Nicolas Bourbaki, epitomizes the clash between the Bourbakites' quest for individual anonymity and their individual egos. Dieudonné once published a paper bearing the name of Bourbaki. The paper was found to contain an error, and so Dieudonné published a note titled: "On an Error of N. Bourbaki," and signed it J. Dieudonné.[11]

The schizophrenic nature of the society—its members were all being French, but most of them lived in the United States—manifests itself in Mr. Bourbaki's affiliation. Usually, publications of Nicolas Bourbaki list his affiliation as "the University of Nancago," a fusion of the name of the French city of Nancy with that of Chicago. But Bourbaki publishes only in French, and when its members meet, usually at a French resort city, the conversations are not only in French, they are in the dialect of the Parisian students. The chauvinism carries into the separate lives of these French mathematicians living in America. André Weil, a founding Bourbakite, published many important papers in English. But his *Collected Works*, which had some bearing on the conjecture related to the Fermat problem, were published in French and titled *Oeuvres*. Weil's unusual actions would hurt one of the principal players in our drama, and Weil would not recover from this debacle.

The members of Bourbaki must be given credit for their collective sense of humor. Some forty years ago, the American Mathematical Society received an application for individual membership from Mr. Nicolas Bourbaki. The secretary of the Society was imperturbable. He replied that if Mr. Bourbaki wanted to join the Society he would have to apply as an institutional member (which was much more expensive). Bourbaki did not write back.

Elliptic Curves

Diophantine problems—that is, problems raised by equations of the form given by Diophantus in the third century—began to be studied more and more in the twentieth century using mathematical entities called elliptic curves. But elliptic curves have nothing to do with ellipses. They were originally used centuries earlier in connection with elliptic functions, which in turn were devised to help calculate the perimeter of an ellipse. As with many innovative ideas in mathematics, the pioneer in this field was none other than Gauss.

Oddly, elliptic curves are neither ellipses nor elliptic functions—they are cubic polynomials in two variables. They look like: $y^2 = ax^3 + bx^2 + cx$ where a, b, and c are integers or rational numbers (we are concerned with elliptic curves over the rational numbers). Examples of such elliptic curves are shown below.[13]

When one looks at the rational points on the elliptic curves—that is, one looks only at points on the curve that are ratios of two integers (no irrational numbers such as pi or the square root of two, etc.), these numbers form a group. That means that they have nice properties. Take any two solutions; they can be "added" in a sense to produce a *third* solution on the curve. Number theorists have become fascinated with the elliptic curves since they can answer many questions about equations and their solutions. Elliptic curves thus became one of the foremost research tools in number theory.[14]

A Strange Conjecture Is about to be Made

It was known for some time by number theory experts that *some* of the elliptic curves they were studying were modular.

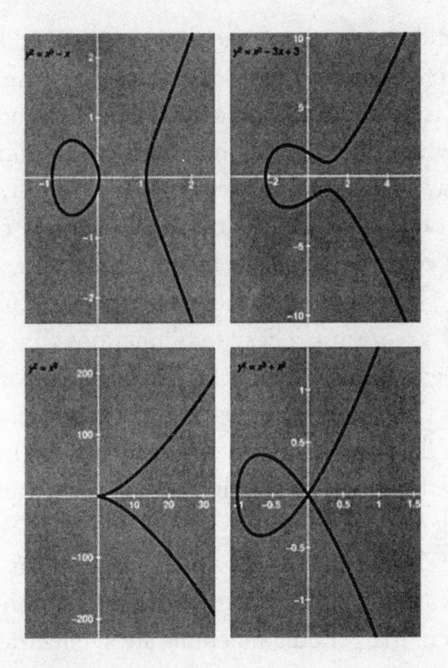

That is, these few elliptic curves could be viewed as connected with modular forms. Some elliptic curves could be somehow connected with the complex plane and these functions in hyperbolic space with their many symmetries. It was not clear why and how this was happening. The mathematics was extremely complicated, even for experts, and the internal structure—the beautiful harmonies that existed—was little understood. The elliptic functions that were indeed modular had interesting properties. Soon someone would make the bold conjecture that *all* elliptic functions were modular.

To understand the idea of modularity, which exists within

the non-Euclidean space of the upper complex half-plane, where symmetries are so complicated that they can hardly be visualized, it is useful to look at a very simple example. This is an example where the curve of interest is not an elliptic curve; instead of a cubic equation in two variables, it is only a squared equation in two variables: the curve is a simple circle. The equation of a circle with radius a^2 whose center is zero is given by: $x^2 + y^2 = a^2$. Now look at the simple periodic functions: $x = a \cos t$, and $y = a \sin t$. These two functions can stand for x and y in the equation of the circle. The equation of the circle is modular in this sense. The reason is the trigonometric identity that says that $\sin^2 t + \cos^2 t = 1$, and substituting this formula into the equation of the circle (each term multiplied by a) gives an identity.

A modular elliptic curve is just an extension of this idea to the more complicated complex plane, with a special non-Euclidean geometry. Here the periodic functions are symmetries not only with respect to one variable, t, as with the sines and cosines on the line—they are the automorphic, or the modular forms on the complex plane, which have symmetries with respect to more complicated transformations ($f(z)$—->$f(az+b/cz+d)$).

Tokyo, Japan, Early 1950s

In the early 1950s, Japan was a nation emerging from the devastation of war. People were no longer hungry, but they were still poor and everyday survival was a struggle for the average Japanese. Yet factories were being rebuilt from the rubble, businesses reestablished, and the general mood was hopeful.

University life in Japan at that time was also difficult. Competition among students was fierce: good grades meant good jobs after graduation. This was especially true for doctoral students in pure mathematics, since positions at universities were scarce even though the pay was low. Yutaka Taniyama was one such doctoral student in mathematics. He was born on November 12, 1927, the youngest of eight children in the family of a country doctor in the town of Kisai, about 30 miles north of Tokyo. At a young age, Taniyama began to study the area of mathematics involving complex multiplication of abelian varieties. Not much was known about this field and Taniyama had a very difficult time. To make things worse, he found the advice of older professors at the University of Tokyo virtually useless. He had to derive every detail on his own and he used to describe every task in his mathematical research using four Chinese characters that meant "hard fighting" and "bitter struggle." Nothing was easy in young Yutaka Taniyama's life.

Taniyama lived in a one-room apartment of 81 square feet. There was only one toilet on every floor of the building, shared by all the residents of the floor. To take a bath, Taniyama had to go to a public bathhouse some distance away from his building. The shabby apartment building was named "Villa Tranquil Mountains," ironically so since it stood on a busy street and by a railroad track on which trains thundered by every few minutes. Possibly so he could concentrate better on his research, young Yutaka worked mostly at night, often going to bed at 6 AM when the noisy day began. Except in the heat of summer, almost every day Taniyama wore the same blue-green suit with a metallic sheen. As he explained to his good friend Goro Shimura, his father bought the material very cheaply from a peddler. But because of the luster, no one in the family dared wear it. Yutaka, who didn't care how he looked, finally volunteered and had the material tailored for a suit which became his daily outfit.

Taniyama graduated from the University of Tokyo in 1953 and took on the position of "special research student" at the department of mathematics there. His friend Shimura graduated a year earlier and had a similar position in mathematics at the College of General Education across campus. Their friendship began after one of them wrote the other a letter asking him to return to the library an issue of a mathematical journal which interested them both. They would often eat together at inexpensive restaurants supposedly serving Western-style food, such as tongue stew, which was becoming popular in Japan.[15]

Few good mathematicians stayed in Japan in those days. Once a mathematician achieved some renown, he or she would try to move to a university in America or Europe, where the mathematical community was more established and where connections with people doing research in the same fields were possible. Such links were important for conducting research in esoteric areas about which not much was known. To try to foster research ties with people with knowledge in their field of interest, the two friends helped organize the Tokyo-Nikko Symposium on Algebraic Number Theory in September, 1955. Some statements made at this small conference, while destined to remain obscure for a long time, would eventually lead to momentous results—and a savage controversy—almost forty years later.

COURTESY GORO SHIMURA

Yutaka Taniyama.

COURTESY GORO SHIMURA

A tram journey to Nikko at the 1955 conference. Left to right: T. Tamagawa, J.-P. Serre, Y. Taniyama, A. Weil.

COURTESY PRINCETON UNIVERSITY

Goro Shimura, circa 1965, when he first developed his conjecture

A Hopeful Beginning

The two friends filed the necessary forms with the administration, arranged for the conference facilities, and sent out invitations to local and foreign mathematicians they hoped would attend the conference. André Weil, who had left France in the meantime and was a professor at the University of Chicago, was one of those invited to attend. At the International Congress of Mathematicians five years earlier, Weil had brought to the attention of the mathematical community an unknown conjecture by a mathematician by the name of Hasse about the "zeta function of a variety over a number field." The obscure statement held some interest to researchers in number theory. Apparently, Weil was collecting these conjectural ideas in the theory of numbers and included this one in his *Collected Papers*, giving credit to Hasse.

His interest in results in this area made him attractive to Taniyama and Shimura, and they were pleased when he accepted the invitation to attend their conference. Another foreign mathematician to come to Tokyo-Nikko was a younger French mathematician, Jean-Pierre Serre. While he may not have been a member of Bourbaki at that time, since the society included only well-known mathematicians, he would become one within the following decades. Serre has been described by some mathematicians as ambitious and fiercely competitive. He came to Tokyo-Nikko to learn as much as he could. The Japanese knew some things about number theory, and they had many results published only in Japanese and thus hidden from the rest of the world. This was a great opportunity to learn of these results, since the conference was to be conducted in English. He would be one of few people outside Japan with knowledge of the mathematics presented there. When the proceedings of the symposium would be published, it would be in Japanese only. Twenty years later, Serre would draw attention to some events at Tokyo-Nikko, and the world would hear his version—not the one recorded in the Japanese proceedings.

The proceedings included thirty-six problems. Problems 10, 11, 12, and 13 were written by Yutaka Taniyama. Similar to the ideas of Hasse, Taniyama's problems constituted a conjecture about zeta functions. He seemed to connect Poincaré's automorphic functions of the complex plane with the zeta function of an elliptic curve. It was mysterious that an elliptic curve should somehow be connected with something in the complex plane.

"You Are Saying What...?"

The conjecture embodied in the four problems was nebulous. Taniyama did not formulate the problems in a very meaningful way, possibly because he wasn't quite sure what the connection was. But the basic idea was there. It was an intuition, a gut feeling that the automorphic functions with their many symmetries on the complex plane were somehow connected with the equations of Diophantus. It certainly wasn't obvious. He was positing a hidden connection between two very different branches of mathematics.

André Weil wanted to know exactly what Taniyama had in mind. According to the written record of the conference, the proceedings published in Japanese, the following exchange took place between Weil and Taniyama:[16]

> Weil: Do you think all elliptic functions are uniformized by modular functions?
>
> Taniyama: Modular functions alone will not be enough. I think other special types of automorphic functions are necessary.
>
> Weil: Of course some of them can probably be handled that way. But in the general case, they look completely different and mysterious...

Two things are evident from the conversation. First, Taniyama was referring to "automorphic functions" rather than "modular functions alone" as being associated with the elliptic curves. And second, Weil did not believe that in general there was such a connection. Later he would be more specific about this disbelief, all of which would make it astounding that *his* name, of all people's, should end up being associated with a conjecture he neither formulated nor even believed was true. But fate sometimes takes strange, implausible turns, and even more bizarre occurrences were going to transpire.

All this would matter decades later. To know exactly what Yutaka Taniyama meant, thought, and said would be a boon to modern historians. But, unfortunately, tragedy stalked Taniyama, as it did so many other young mathematical geniuses.

Within a couple of years, Goro Shimura left Tokyo, first for Paris, then for the Institute for Advanced Study and Princeton University. The two friends continued to communicate by mail. In September, 1958, Goro Shimura received the last letter written by Yutaka Taniyama. In the morning of November 17, 1958, five days after his thirty-first birthday, Yutaka Taniyama was found dead in his apartment, a suicide note on his desk.

Shimura's Conjecture

A decade passed since the Tokyo-Nikko conference and Goro Shimura, now at Princeton, continued his research on number theory, zeta functions, and elliptic curves. He understood where his late friend had been wrong, and his own research and quest for hidden harmonies in the realms of mathematics led him to formulate a different, bolder and more precise conjecture. His conjecture was that every elliptic curve over the rational numbers is uniformized by a modular form. Modular forms are more specific elements over the complex plane than are the automorphic functions of Taniyama. And specifying the domain as the rational numbers, and other modifications, were important corrections as well.

Shimura's conjecture can be explained using a picture:

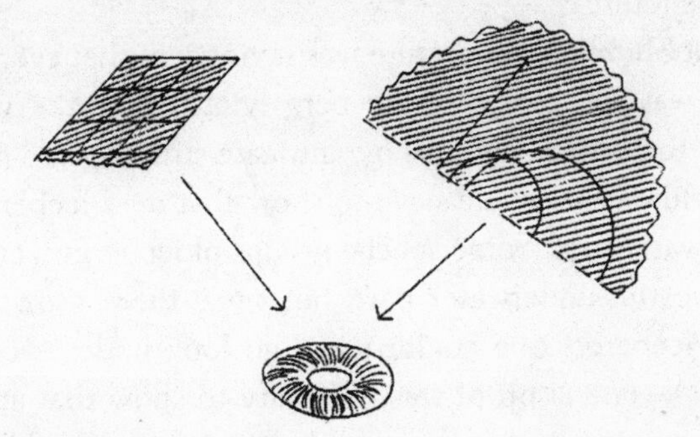

If we "fold" the complex plane as a torus (the doughnut in the picture), then this surface will hold all solutions to elliptic equations over the rational numbers, these in turn arising from the equations of Diophantus. What would later be important to the solution of Fermat's Last Theorem is that if a solution to Fermat's equation $x^n + y^n = z^n$ existed, this solution would also have to lie on that torus. Now, Shimura conjectured that every elliptic curve with rational coefficients (that is, an equation with coefficients of the form *a*/*b* where both *a* and *b* are integers) has a "mate" on the complex half-plane of Poincaré, with its non-Euclidean, hyperbolic geometry. The particular mate of each rational elliptic curve was a very specific function on the complex half-plane, which was invariant under complicated transformations of the plane—the ones mentioned earlier:

$f(z) \longrightarrow f(az+b/cz+d)$, the coefficients forming a group with many unexpected symmetries. All of this was very complex, very technical, and—as most mathematicians would believe for several decades—impossible to prove in the foreseeable future.

What Shimura's conjecture was saying was that every elliptic curve was the part of an iceberg lying above the waterline. Below the surface lay a whole intricate structure. To prove the conjecture, one would have to show that *every* iceberg had an underwater part. Some special groups of icebergs were known to have the underwater part, but since there were infinitely many icebergs, one couldn't just go look under each one of them. A general proof was necessary to show that an iceberg couldn't exist without part of it being underwater. The formulation of such a proof was considered exceedingly difficult.

Intrigue and a Betrayal

At a party at the Institute for Advanced Study at Princeton in the early 1960s, Shimura again met Jean-Pierre Serre. According to Shimura, Serre approached him rather arrogantly. "I don't think that your results on modular curves are any good," he said. "Why, they don't even apply to an arbitrary elliptic curve." In response, Shimura stated his conjecture exactly: "Such a curve should *always* be uniformized by a modular curve."[17] Serre went to Weil, who was not at the party but was a member of the Institute and therefore in the immediate area, and told him of his conversation with Shimura. In response, André Weil came to

Shimura. "Did you really say that?" he asked him, puzzled. "Yes," answered Shimura, "don't you think it plausible?" Ten years after Taniyama's related conjecture, André Weil still did not believe either conjecture. He answered: "I don't see any reason against it, since one and the other of these sets are denumerable, but I don't see any reason either for this hypothesis." What Weil said on this occasion would later be described as "stupid," and "inane" by Serge Lang of Yale University, who would circulate these comments together with copies of two dozen letters he named collectively "The Taniyama-Shimura File," among about fifty mathematicians worldwide. What Weil meant in his response to Shimura was tantamount to the following: If in a room you have seven men and seven women and you conjecture that these are seven married couples, then I see no reason against it, since the number of men is the same as the number of women. But I don't see any reason for your hypothesis, either. It could be that they are all single. What made Lang describe the statement as stupid was that the counting argument didn't really apply in any simple way here, because "denumerable" means infinite and countable (such as the number of all the positive integers: 1, 2, 3, 4,...) and matching such infinite collections is no simple task. At any rate, it is clear that André Weil did not believe Shimura's theory was necessarily true. He would later admit that the conversation took place and, stupid, inane, or otherwise, would quote it. But this would happen only in 1979, when he would write:[18]

> Quelques années plus tard, à Princeton, Shimura me demanda si je trouvais plausible que toute courbe elliptique sur Q fut contenue dans le jacobienne d'une courbe definie par une sous-groupe de

> congruence du groupe modulaire; je lui repondis, il me semble, que je n'y voyais pas d'empêchement, puisque l'un et l'autre ensemble est dénombrable, mais je ne voyais rien non plus qui parlat en faveur de cette hypothèse.

["Some years later, at Princeton, Shimura asked me if I found it plausible that every elliptic curve over Q was continued in the jacobian of a curve defined by a congruence subgroup of a modular group; I responded to him, it seems to me, that I don't see anything against it, since one set and the other are denumerable, but neither do I see anything that speaks in favor of this hypothesis."]

But even then, Weil would write "Shimura asked me" (me *demanda*), rather than "Shimura told me," when referring to the statement that is Shimura's conjecture. Weil published some related papers, and while he did not believe Shimura's theory, his own name became associated with it. The error was perpetuated when mathematicians made references in their papers to the works of others, and the misquotation is present to this day when writers ignorant of the history refer to the Weil-Taniyama conjecture instead of the Shimura-Taniyama conjecture. Weil seemed to enjoy his association with an important theory which—while he himself did not believe in it—most mathematicians thought would be proved some day in the distant future.

With the passing decades, there was more and more reason for the connection to exist. If and when the conjecture was proved, it would be a substantial mathematical theory. Weil worked around the conjecture, never leaving mathematical results he obtained too far away from the possible connection between modular forms of the complex plane and the elliptic curves of Diophantine equations. And while he certainly knew better, he held back references to Shimura and his crucial role until almost two decades had passed. Then he gave offhand praise to Shimura in casual conversation and mentioned him—almost in passing—in a published paper. Meanwhile, in France, Serre was actively contributing to the false attribution. He made every effort to associate the name of André Weil with the conjecture, instead of that of Goro Shimura.

"An Exercise for the Interested Reader"

In 1967, André Weil wrote a paper in German, in which he said:[19]

> Ob sich diese Dinge immer, d.h. für jede über Q definierte Kurve C, so verhalten, scheint im Moment noch problematisch zu sein und mag dem interessierten Leser als Übungsaufgabe empfohlen werden.

["Whether these things, that is for every curve C defined over Q, so behave, at this moment is still seen as problematic and will be recommended as an exercise for the interested reader."]

This paragraph refers to elliptic curves over the rational numbers (which mathematicians denote by Q), and "sich so verhalten" here refers to being modular, that is, it states Shimura's conjecture. But, here again, Weil did not attribute the theory to its originator. He did so only 12 years later, and even then as "Shimura *asked* me..." as we have just seen. In this paper in German, above, Weil calls the conjecture "problematic." And then he does something strange. He simply assigns the conjecture as an *exercise for the interested reader* ("und mag dem interessierten Leser als Übungsaufgabe empfohlen werden"). This exercise for the "interested reader" would take one of the world's finest mathematicians seven years of work in solitude to attempt to prove. When a mathematician assigns a homework problem (*Übungsaufgabe*), usually he or she knows the proof through and through, and believes—knows for certain—that the theorem is true, not "problematic" as Weil describes it.

There is an old story about a math professor who tells his class "this is obvious," when referring to some concept. The class looks confused since it is not at all obvious, and finally a bold student raises a hand and asks, "Why?" The professor then starts drawing and writing on the edge of the board with one hand, covering the writing with his other hand, and erasing everything as he is done. After about ten minutes of this furtive scribbling, the professor erases the board completely and announces to the befuddled class: "Yes, it's obvious."

The Lie

In the 1970s, Taniyama's problems from the Tokyo-Nikko meeting received wider distribution. In the meantime, since Weil had written about the conjecture he doubted, modular elliptic curves became known as "Weil Curves." When Taniyama's problems became better known in the West, the conjecture about such curves came to be called the "Taniyama-Weil conjecture." Shimura's name was left out completely. But since Taniyama's name came in, Weil started to inveigh against conjectures altogether. In 1979, a mere five years before it was proved by Gerd Faltings, Weil even spoke against "the so-called 'Mordell conjecture' on Diophantine equations." He continued, "It would be nice if this were so, and I would rather bet for it then against it. But it is no more than wishful thinking because there is not a shred of evidence for it, and also none against it." But Weil was wrong even then. A number of Russian mathematicians, among them Shafarevich and Parshin, were already obtaining results that would supply evidence for the Mordell conjecture as early as the early 1970s. In 1984, of course, Gerd Faltings would prove the conjecture outright, making Fermat's Last Theorem "almost always true."

While André Weil was turning against all conjectures at a time when his name was no longer being exclusively associated with the conjecture now called Taniyama-Weil by many mathematicians, Serre in Paris was working to keep Shimura's name dissociated from the conjecture. In 1986, at a party at the University of California at Berkeley, and within earshot of a number of people, Jean-Pierre Serre told Serge Lang that André Weil had told him of a conversation he had had with Goro Shimura. According to Serre, this is what Weil told him took place:

> Weil: Why did Taniyama think that all elliptic curves are modular?
> Shimura: You told him so yourself, and you have forgotten.

At this moment, Lang, who unknowingly had been using the terms "Weil curve" and "Taniyama-Weil conjecture," became suspicious. He took it upon himself to find the truth. Lang immediately wrote to both Weil and Shimura, then to Serre. Shimura categorically denied that such a conversation ever took place, and gave ample evidence for this claim. Weil did not reply right away. And Serre, in his response, criticized Lang's attempt to find the truth. In his Bourbaki Seminar in June 1995, Serre still referred to the conjecture as that of "Taniyama-Weil," leaving out the name of its originator, who trusted him with his conjecture 30 years earlier. Weil responded after a second attempt to contact him by Lang. His letter follows.[20]

3 December 1986.

Dear Lang,

I do not recall when and where your letter of August 9 first reached me. When it did, I had (and still have) far more serious matters to think about.

I cannot but resent strongly any suggestion that I ever sought to diminish the credit due to Taniyama and to Shimura. I am glad to see that you admire them. So do I.

Reports of conversations held long ago are open to misunderstandings. You choose to regard them as "history"; they are not. At best they are anecdotes. Concerning the controversy which you have found fit to raise, Shimura's letters seem to me to put an end to it, once and for all.

As to attaching names to concepts, theorems, or (?) conjectures, I have often said: (a) that, when a proper name gets attached to (say) a concept, this should never be taken as a sign that the author in question had anything to do with the concept; more often than not, the opposite is true. Pythagoras had nothing to do with "his" theorem, nor Fuchs with the Fonctions fuchsiennes, any more than Auguste Comte with rue Auguste-Comte; (b) proper names tend, quite properly, to get replaced by more appropriate ones; the Leray-Koszul sequence is now a spectral sequence (and as Siegel once told Erdös, abelian is now written with a small a).

Why shouldn't I have made "stupid" remarks sometimes, as you are pleased to say? But indeed, I was "out of it" in 1979 when expressing some skepticism about Mordell's conjecture, since at that time I was totally ignorant of the work of the Russians (Parshin, etc.) in that direction. My excuse, if it is one, is that I had had long conversations with Shafarevich in 1972, and he never mentioned any of that work.

Sincerely,
A. Weil
AW:ig

P.S. Should you wish to run this letter through your Xerox machine, do feel free to do so. I wonder what the Xerox Co. would do without you and the like of you.

Deep in the Black Forest, Fall 1984

While the controversy about who originated the Shimura-Taniyama conjecture was raging in Berkeley, New Haven, Princeton, and across the Atlantic in Paris, something totally unexpected was happening deep in the Black Forest of southwest Germany.

Gerhard Frey received his Diploma from the University of Tübingen, and his Ph.D. from the University of Heidelberg, where he studied number theory and was influenced by the works of Hasse and Weil. Frey was fascinated by the interplay between the theory of numbers and algebraic geometry, an area of mathematics which was developed in the last fifty years. He was also interested in arithmetic geometry. It was the connections between number theory and the newer fields of algebraic and arithmetic geometry that would lead him to make an unexpected mathematical statement. In the 1970s, Frey did a lot of work on elliptic curves and Diophantine equations, and in 1978 he read the paper "Modular curves and the Eisenstein ideal," by Barry Mazur of Harvard University. Frey was strongly influenced by the paper, as were many number theorists, among them Berkeley's Kenneth Ribet and Princeton's Andrew Wiles. Frey became convinced by Mazur's paper that he should think very seriously about applications of modular curves and Galois representations to the theory of elliptic curves. He found that this led him almost unavoidably to Diophantine questions closely related to equations of Fermat's type. This was a powerful insight, which Frey tried to make more precise.

In 1981, Gerhard Frey spent a few weeks at Harvard University and had a number of discussions with Barry Mazur. These discussions were clearing things in his mind. The heavy fog surrounding the difficult connections he envisioned between Fermat-like equations and the relation between modular forms and elliptic curves was slowly lifting. Frey went on to Berkeley, where he spoke with Ken Ribet, a bright number theorist who was a graduate of Harvard and had worked with Mazur on related issues. Frey returned to his native Germany.

Three years later, he was invited to give a lecture at the Oberwolfach center deep in the Black Forest.

Oberwolfach was designed as a conference and workshop center in mathematics, set in beautiful and peaceful surroundings far from cities and crowds. Every year, the center sponsors about fifty international meetings on different topics of mathematics. Lectures, and even just attendance at the meetings, are exclusively by invitation. Every effort is made to allow for the easy exchange of ideas among experts from different countries. In 1984, Gerhard Frey gave a talk at a number theory conference there. He made what looked like a crazy assertion. The mimeographed sheets filled with mathematical formulas he passed around at the conference seemed to imply that if the Shimura-Taniyama conjecture were indeed true, Fermat's Last Theorem would be proved. This made no sense at all. When Ken Ribet first heard of Frey's statement, he thought it was a joke. What could modularity of elliptic curves possibly have to do with Fermat's Last Theorem? he asked himself. He gave this strange assertion no further thought and went about his usual work. But a number of people in Paris and elsewhere were interested in Frey's unproven, and somewhat incomplete statement. Jean-Pierre Serre wrote a private letter to a mathematician by the name of J.-F. Mestre. This letter became public, and Serre subsequently published a paper repeating his own conjectures from the letter to Mestre.[21]

Ribet's Theorem

Ken Ribet, who first thought this statement was a joke, started thinking about Serre's conjectures, and in fact recognized in them something he had already formulated for himself when he found time to think about Frey's "joke." These were certain clarifications of Gerhard Frey's statements, which, if proven, would establish the following implication:

Shimura-Taniyama conjecture ————-> Fermat's Last Theorem

The way the Frey idea worked was ingenious. Frey reasoned as follows: Suppose that Fermat's Last Theorem is *not* true. Then, for some power n that is greater than 2 there *is* a solution to Fermat's equation: $x^n + y^n = z^n$, where x, y, and z are integers. This particular solution, a, b, and c, would then give rise to a specific elliptic curve. Now Frey wrote down the general equation of this curve that would result from the solution of Fermat's equation. His conjecture presented at Oberwolfach stated that this very curve, now called the Frey curve, was a very strange animal. It was so strange, in fact, that it couldn't possibly exist. And, most important, the elliptic curve that would arise if Fermat's Last Theorem were false was definitely *not* modular. So, if the Shimura-Taniyama conjecture was indeed true, then all elliptic curves must be modular. Therefore, an elliptic curve that was not modular couldn't possibly exist. And it would follow that Frey's curve, an elliptic curve that was not modular (in addition to all its other strange characteristics) could not exist. Therefore, the solutions to Fermat's equation could not exist either. Without the existence of solutions to the Fermat equation, Fermat's Last Theorem (which states that there are *no* solutions to the equation for any $n>2$), would be proved. This was a complicated sequence of implications, but it followed beautifully the logic of mathematical

COURTESY C. J. MOZZOCHI

Ken Ribet presenting his important theorems.

COURTESY C. J. MOZZOCHI

Andrew Wiles being interviewed.

COURTESY C. J. MOZZOCHI

Barry Mazur of Harvard University—the "granddaddy" of them all, a mathematician whose geometrical insights inspired everyone who contributed to the final proof of Fermat's Last Theorem.

COURTESY C. J. MOZZOCHI

Gerhard Frey, who had the "crazy idea" that an elliptic curve resulting from a solution of Fermat's equation simply could not exist.

proof. The logic was: A implies B; therefore, if B is *not* true, then A can not be true either. However, the Frey statement was, itself, a *conjecture*. It was a conjecture which said that if another conjecture (Shimura-Taniyama) was true, then Fermat's Last Theorem would be established. The pair of subsequent conjectures in Serre's letter to Mestre further allowed Ken Ribet to think about the Frey conjecture in clear terms.

Ken Ribet had never before been interested in Fermat's Last Theorem. He had started out as a chemistry major at Brown University. Under the influence and tutelage of Kenneth F. Ireland, Ribet was steered to mathematics and got interested in zeta functions, exponential sums, and number theory. He had dismissed Fermat's Last Theorem as "one of those problems about which nothing further of real importance could be said." This was a view held by many mathematicians, because problems in number theory tend to be isolated, with no unifying scheme or underlying general principle behind them. What is interesting about Fermat's Last Theorem, however, is that it spans mathematical history from the dawn of civilization to our own time. And the theorem's ultimate solution also spans the breadth of mathematics, involving fields other than number theory: algebra, analysis, geometry, and topology—virtually all of mathematics.

Ribet went on to pursue a Ph.D. in mathematics at Harvard University. There, first indirectly and toward his graduation more directly, he fell under the influence of the great number theorist and geometer Barry Mazur, whose vision inspired every mathematician involved even in the smallest way in efforts to prove Fermat's Last Theorem. Mazur's paper on the

Eisenstein ideal acted as an abstraction of the theory of ideals developed in the last century by Ernst Kummer into the modern fields of mathematics, algebraic geometry and new approaches to the theory of numbers through geometry.[22]

Ken Ribet eventually became a professor of mathematics at the University of California at Berkeley and did research in number theory. In 1985, he heard about Frey's "crazy" notion that if a solution of Fermat's equation existed, that is, if Fermat's Last Theorem were false, then this solution would give rise to a very weird curve. This Frey Curve would be associated with an elliptic curve that could not be modular. The pair of associated conjectures in Serre's letter to Mestre made him interested in trying to prove Frey's conjecture. While he wasn't really interested in Fermat's Last Theorem, Ribet recognized that this had become a hot problem, and it happened to be in an area he knew well. During the week of August 18–24, 1985, Ribet was at a meeting on arithmetic algebraic geometry in Arcata, California. He began thinking about Frey's statement, and the problem remained on his mind for the next year. When he was freed of his teaching obligations at Berkeley early in the summer of 1986, Ribet flew to Germany where he was to do research at the world-famous center for mathematics, the Max Planck Institute. Just as he arrived at the Institute, Ribet made his great breakthrough. He was now almost able to prove Frey's conjecture.

But he was not quite there. When he returned to Berkeley, Ribet ran into Barry Mazur, who was visiting from Harvard. "Barry, let's go for a cup of coffee," Ribet suggested. The two retreated to a popular café by the University of California campus. While sipping a cappuccino, Ribet confided to

Mazur: "I'm trying to generalize what I've done, so that I'll be able to prove the Frey conjecture. I just don't seem to get this one thing to generalize it..." Mazur looked at what he was showing him. "But you've done it already, Ken," he said, "all you need to do is add on some extra gamma zero of N structure, and run through your argument, and you're there!" Ribet looked at Mazur, he looked back at his cappuccino, then back at Mazur with disbelief. "My God, you're absolutely right!" he said. Later he went back to his office to finish off the proof. "Ken's idea was brilliant," Mazur exclaimed when describing Ken Ribet's ingenious proof after it was published and became known in the world of mathematicians.

Ribet formulated and proved a theorem which established as fact that if the Shimura-Taniyama conjecture was true, then Fermat's Last Theorem would fall out of it as a direct consequence. The man who only a year earlier thought that Frey's suggestion was a joke now proved that the "joke" was actually a mathematical reality. The door to the attack on Fermat's problem using the modern methods of arithmetic algebraic geometry was now wide open. All the world needed now was someone who would prove the seemingly-impossible Shimura-Taniyama conjecture. Then Fermat's Last Theorem would automatically be true.

A Child's Dream

The person who wanted do just that was Andrew Wiles. When he was ten years old, Andrew Wiles went to the public library in his town in England and looked at a book on mathematics. In that book he read about Fermat's Last Theorem. The theorem, as described in the book, seemed so simple, that any child could understand it. In Wiles' own words: "It said that you will never find numbers, x, y, and z, so that $x^3 + y^3 = z^3$. No matter how hard you tried, you will never, ever find such numbers. And it said that the same was true for $x^4 + y^4 = z^4$, and for $x^5 + y^5 = z^5$, and so on... It seemed so simple. And it said that nobody has ever found a proof of this for over three hundred years. I wanted to prove it..."

In the 1970s, Andrew Wiles went to the university. When he finished his degree he was admitted as a research student in mathematics to Cambridge. His adviser was Professor John Coates. Wiles had to drop his childhood dream of proving Fermat's Last Theorem. Research on the problem would have turned into such a waste of time that no graduate student could afford it. Besides, what doctoral adviser would have accepted a student working on such an ancient puzzle, one that had kept the world's brightest minds from a solution for three centuries? In the 1970s, Fermat was not in fashion. What was "in" at the time, the real hot topic for research in number theory, was elliptic curves. So Andrew Wiles spent his time doing research on elliptic curves and in an area called Iwasawa theory. He completed his doctoral dissertation, and when he was awarded his Ph.D., he got a position in mathematics at Princeton University and moved to the United States. There, he continued doing research on elliptic curves and Iwasawa theory.

An Old Flame Rekindled

It was a warm summer evening, and Andrew Wiles was sipping iced tea at a friend's house. Suddenly, in the middle of the conversation, his friend said: "By the way, did you hear that Ken

Ribet just proved the Epsilon Conjecture?" The Epsilon Conjecture is what number theorists were informally calling Frey's conjecture, as modified by Serre, about the connection between Fermat's Last Theorem and the Shimura-Taniyama Conjecture. Wiles was electrified. At that moment, he knew that his life was changing. The childhood dream he had of proving Fermat's Last Theorem—a dream he had had to give up to undertake more feasible research—came alive again with incredible force. He went home and started thinking about how he would prove the Shimura-Taniyama conjecture.

"For the first few years," he later confided, "I knew I had no competition, since I knew that nobody—me included—had any idea where to start." He decided to work in complete secrecy, and in isolation. "Too many spectators would spoil the concentration, and I discovered early on that just a mention of Fermat immediately generates too much interest." Of course, gifted, able mathematicians abound, especially at a place like Princeton, and the danger of someone completing your work for you—and even doing it better—is very real.

Whatever the reason, Wiles locked himself up in his attic office and went to work. He abandoned all other research projects to devote his time completely to Fermat. Wiles would use all the power of the modern machinery of algebra, geometry, analysis, and the other areas of mathematics. He would also make use of the important mathematical results of his contemporaries, and of his historical predecessors. He would make use of Ribet's clever methods of proof, and his results; he would use the theories of Barry Mazur, and the ideas of Shimura, Frey, Serre, André Weil, and those of many, many other mathematicians.

Wiles' greatness, Gerhard Frey would later say, was that he believed in what he was doing at a time when virtually every mathematician in the world believed that the Shimura-Taniyama conjecture could not be proven in the twentieth century.

To prove the Shimura-Taniyama Conjecture, Andrew Wiles knew he had to prove that every elliptic curve is modular. He had to show that every elliptic curve, whose solutions lie on a doughnut, was really a modular form in disguise. The doughnut was, in a sense, also this space of intricately symmetric functions on the complex plane called modular forms. Nobody had any idea how to show such a weird connection between these two seemingly very different entities.

Wiles realized that the best idea was to try to *count* the number of elliptic curves, and to count the number of *modular* elliptic curves, and then to show that their "number" was the same. This construction would prove that the elliptic curves and the modular elliptic curves were one and the same, and hence every elliptic curve is indeed modular, as the Shimura-Taniyama conjecture claims.

Wiles realized two things. One was that he didn't have to prove the entire Shimura-Taniyama conjecture, but only a special case: *semistable* elliptic curves with rational numbers as the coefficients. Proving that the conjecture was true for this smaller class of elliptic curves would be enough to establish Fermat's Last Theorem. The other thing Wiles knew was that "counting" would not work here because he was dealing with *infinite* sets. The set of semistable elliptic curves was infinite. Any rational number a/b, where a and b are integers, would give you another elliptic curve (we say an elliptic curve *over* the rationals). Since there are infinitely many such numbers—*a* and *b* can be any of the infinitely many numbers 1,2,3,4..., to infinity, there are infinitely many elliptic curves. So counting as we know it wouldn't work.

Breaking Down a Big Problem into Smaller Ones

Wiles thought he might try to work on smaller problems, one at a time. Maybe he could look at sets of elliptic functions and see what he could do about them. This was a good approach since it broke down the task so that, step by step, he could understand each set. First of all, some elliptic curves were already known to be modular. These were very important results, developed by many other number theorists. But soon Andrew Wiles realized that looking only at elliptic curves and trying to count them off against modular forms might not be a good approach—he was dealing with two infinite sets. In fact, he was no closer to a solution than was the skeptical André Weil when he said: "I see no reason against the conjecture since one and the other of the two sets are denumerable [infinite but of the order of infinity of the integers and rational numbers, not the higher order of infinity of the irrational numbers and the continuum], but I don't see any reason for it, either..." After two years of getting nowhere, Wiles tried a new approach. He thought he might *transform* the elliptic curves into Galois representations, and then count these Galois representations against the modular forms.

The idea was an excellent one, although it was not original. The principle behind this move is interesting. Number theorists are concerned with finding solutions of equations, such as the Fermat equation. The mathematical theory of fields of numbers sets this problem in the context of field extensions. Fields are large, infinite collections which are difficult to analyze. Therefore, what number theorists have often done is to use the theories of Évariste Galois, called Galois theory, in order to translate these problems from the complicated fields to what are known as groups. Often a group is generated by a finite (rather than an infinite) set of elements. Using Galois theory thus allows the number theorist to move from an infinite collection to one that is represented by a finite set. This translation of a problem constitutes an immense step forward, since a finite set of elements is so much easier to handle than an infinite set. Counting makes sense only for a finite number of elements. The approach seemed to work for some sets of elliptic curves. This was a good breakthrough. But after another year, Wiles was stuck again.

The Flach Paper

What Andrew Wiles was trying to do now was to count sets of Galois representations, corresponding to the (semi-stable) elliptic curves against the modular forms, and to show that they are the same. In doing so, he was using his area of expertise, in which he had done his dissertation, called Horizontal Iwasawa Theory. Wiles was trying to use this theory to attain the Class Number Formula, a result which he needed for the "counting." But here he was up against a brick wall. Nothing he could do brought him nearer to the answer.

In the summer of 1991, Wiles was at a conference in Boston, where he met his former doctoral adviser at Cambridge, John

Coates. Professor Coates told Wiles that one of Coates' students, Matthias Flach, using earlier work by a Russian mathematician named Kolyvagin, had devised an Euler System (named after Leonhard Euler) in an attempt to prove the Class Number Formula. This was exactly what Wiles needed for his proof of the Shimura-Taniyama conjecture—if, indeed, he could extend Flach's partial results to the full Class Number Formula. Wiles was elated on hearing from Coates about this work of Flach. "This was tailor-made" for his own problem, Wiles said, as if Matthias Flach had done all this work just for him. And Wiles immediately abandoned all his Horizontal Iwasawa Theory work and immersed himself day and night in the work of Kolyvagin and Flach. If their "Euler System" really worked, Wiles would hopefully achieve the Class Number result and the Shimura-Taniyama Conjecture would be proved for semistable elliptic curves—enough to prove Fermat's Last Theorem.

This was hard work, however, and it was outside the Iwasawa realm which Wiles knew so well. Increasingly, Wiles started to feel the need to find somebody to talk to. He wanted someone who could check his progress in these uncharted waters, but someone who would not reveal a thing to anyone else.

A Good Friend

Wiles finally had to make a decision: should he continue to keep everything secret as he had done for so long, or should he break down and talk to someone with good knowledge of number theory? He finally decided he could probably not do very well by keeping secrecy forever. As he himself said, one could work on a problem for an entire lifetime and not see any results. The need to compare notes with another person finally outweighed that intense need to keep it all to himself. But now the question was: who? Who could he trust to keep his secret?

In January of 1993, after six years of working alone, Wiles made his contact. He called in Professor Nick Katz, one of his colleagues at the Princeton mathematics department. Katz was an expert on many of the theories that went into attempts to prove the Class Number Formula. But more importantly, Katz was completely trustworthy. He would never reveal what Andrew Wiles was up to. This assessment on Wiles' part turned out to be correct. Nick Katz kept his mouth shut throughout his work with Wiles over the many remaining months of the project. Their mutual colleagues at the tightly-knit mathematical community at Princeton never suspected a thing, even after weeks of seeing the two spend hours huddled together over coffee at the far end of the Commons Room.

But Andrew Wiles still worried that someone might suspect what he was working on. He couldn't take a chance. So he came up with a scheme to hide the fact that he was working very intensely on "something" with Nick Katz. Wiles would offer a new graduate course in mathematics for the spring of 1993, a course which Nick Katz would attend as one of the students, and this would allow the two of them to work together without others suspecting what they were doing. Or at least this is what Wiles has said. The graduate students could not suspect that behind these lectures was a road to Fermat's Last Theorem, and Wiles would be able to pick their brains for any possible holes in his theory, with the help of his good friend Katz.

The course was announced. It was called "Calculations with Elliptic Curves," which was innocent enough so no one could suspect anything. And at the start of the course, Professor Wiles said that the purpose of the lectures was to study some recent work of Matthias Flach on the Class Number Formula. There was no mention of Fermat, no mention of Shimura or Taniyama, and no one could suspect that the Class Number Formula they were going to study would be the keystone to proving Fermat's Last Theorem. And no one had any idea that the true purpose of the lectures was not to teach graduate students mathematics but to allow Wiles and Katz to work together on this problem without suspicion by any of their colleagues, while at the same time getting unsuspecting graduate students to check the work for them.

But within a few weeks all the graduate students drifted away. They couldn't keep up with a course that wasn't really going anywhere. The only "student" who seemed to know anything and to participate in class was the other math professor who was sitting there with them. So after a while, Nick Katz was the only one in the audience. But Wiles just went on using the "class" to write his long proof of the Class Number Theorem on the board, continuing to the next step every class meeting, with Nick Katz verifying each step.

The lectures revealed no errors. It seemed that the Class Number Formula was working, and Wiles was on his way to the solution of the Fermat problem. And so in the late spring of 1993, as the course was coming to an end, Andrew Wiles was almost finished. Still, he was wrestling with just one final obstacle. He was able to prove that most of the elliptic curves were modular, but a few of them remained unprovable. He thought he could overcome these difficulties, and he was generally optimistic. Wiles felt it was time to talk to one more person, to try to gain a little more insight into this last difficulty facing him. So he called in another one of his colleagues at the Princeton mathematics department, Professor Peter Sarnak, and swore him to secrecy as well. "I think I am about to prove Fermat's Last Theorem," he told the stunned Sarnak.

"This was incredible," Sarnak later recalled. "I was flabbergasted, elated, disturbed—I mean...I remember finding it difficult to sleep that night." So now there were two colleagues trying to help Wiles finish his proof. While nobody suspected what it was that they were doing, people were noticing something. And while he maintained that no one ever found anything through him, Sarnak later admitted that he dropped "maybe a few hints..."

The Last Piece of the Puzzle

In May, 1993, Andrew Wiles was sitting alone at his desk. He was getting somewhat frustrated. It seemed that those few elliptic curves that got away from him were not coming any nearer. He simply couldn't prove that they were modular. And he needed them, too, if he were to prove that *all* (semistable) elliptic curves were modular so that Fermat's Last Theorem would follow. Doing it for most of the semistable elliptic curves was a great mathematical result on its own right, but not enough to reach his goal. To rest a little from his intense search leading nowhere, Wiles picked up an old paper of the great master, Barry Mazur of Harvard University. Mazur had made some groundbreaking discoveries in number theory—results that had inspired many of the experts in the field, including Ribet and Frey, whose work paved the way for Wiles' effort. Mazur's paper that Wiles was now rereading was an extension of the theory of ideals, starting with Kummer and Dedekind, and continuing with yet a third nineteenth century mathematician—Ferdinand Gotthold Eisenstein (1823–1852). Although he died young, Eisenstein made great strides in number theory. Gauss, in fact, is quoted as having said: "There have been only three epoch-making mathematicians, Archimedes, Newton, and Eisenstein."

Mazur's paper on the Eisenstein Ideal had one line in it that now caught Wiles' attention. Mazur was saying that it was possible to *switch* from one set of elliptic curves to another. The switch had to do with prime numbers. What Mazur was saying was that if one was dealing with elliptic curves that were based on the prime number three, it was possible to transform them so that one could study them using the prime number five instead. This 3-to-5 switch was exactly what Wiles would need. He was stuck with not being able to prove that certain classes of elliptic curves based on the prime number three were modular. And here Mazur was saying that he could switch them to curves based on the prime number five. But these curves based on five Wiles had already proved were modular. So the 3-to-5 switch was the final trick. It took the difficult elliptic curves based on three, transformed them into base five and these were known to be modular. Once again, some other mathematician's brilliant idea helped Wiles overcome a seemingly insurmountable hurdle. Andrew Wiles was finally done.

His timing was perfect, too. In the next month, June, his

former adviser John Coates would be hosting a number theory conference in Cambridge. All the big names in number theory would be there. And Cambridge was Wiles' old home town and where he went to graduate school. Wouldn't it be perfect to present his proof of Fermat's Last Theorem there? Wiles was now racing against the clock. He had to put together his entire proof that the Shimura-Taniyama conjecture was true for semistable elliptic curves. This meant that the Frey curve could not exist. And if the Frey curve could not exist, that meant that solutions to Fermat's equation could not exist for $n>2$ and therefore Fermat's Last Theorem was proven. The write-up took Andrew Wiles 200 pages. He was done just in time to catch his plane for England. And at the end of his last lecture there, he emerged victorious to the sweeping applause, the flashing cameras, and the reporters.

The Aftermath

Now it was time for the peer-review. Usually, a mathematical result—or any academic finding, for that matter—is submitted to a "refereed journal." Such refereed journals are the standard vehicle for scholars to submit their work for possible publication. The journal's charge is then to send the proposed paper to other experts in the appropriate area of study who review the paper's content and determine whether it is correct, and whether the paper makes a contribution worthy of publication. Refereed journal publications are the bread and butter of academia. Tenure and promotion, and often salary levels and raises, are all dependent on a researcher's output of refereed journal articles.

But Andrew Wiles chose a different approach. Instead of submitting his proof to a professional mathematics journal—as almost anyone else would have done—he presented it at a conference. The reason was probably twofold. Throughout the years of work on the proof, Wiles was obsessed with secrecy. If he submitted the proof to a journal, the proof would have been sent to a number of referees chosen by the journal and one of them, or the editors, might have said something to the world at large. Wiles probably was also worried that someone who read the proposed proof might somehow steal it and send it out under his or her own name. This, unfortunately, does happen in academia. The other reason, linked with the first, was that Wiles wanted to maintain the buildup of suspense as he presented his proof at Cambridge.

But even so, having presented the results at a conference, the work would still have to be refereed. The steps would still have to be peer-reviewed, that is, other experts in number theory would have to go through Wiles' proof, line by line, to ascertain that he had indeed established what he set out to prove.

The Deep Gulf Materializes

Wiles' 200-page paper was sent to a number of leading experts in number theory. Some of them quickly expressed concerns, but generally mathematicians thought the proof was probably correct. They would wait to hear the experts' verdict, however. "Oh yes!" said Ken Ribet when I asked him if he believed Wiles' proof. "I was unable to see what some people were saying soon after they read the proof—namely that there was no Euler System here."

One of the experts chosen to go over Wiles' proof was his Princeton friend, Nick Katz. Professor Katz spent two whole months, July and August of 1993, doing nothing but studying the entire proof. Every day, he would sit at his desk and slowly read every line, every mathematical symbol, every logical implication, to make sure that it made perfect sense and that it would indeed be acceptable to any mathematician who might read the proof. Once or twice a day, Katz would send an e-mail message to Andrew Wiles, who stayed away from Princeton that summer, asking him: "What do you mean in this line on this page?" or "I don't see how this implication follows from the one above," etc. In response, Wiles would e-mail back, and if the problem required more details, he would fax the answer to Katz.

One day, when Katz was about two thirds of the way through Wiles' long manuscript, he came across a problem. It seemed innocent enough at first, one of many Wiles had answered earlier to Katz's complete satisfaction. But not this time. In response to Katz's questions, Wiles e-mailed back an answer. But Katz had to e-mail back: "I still don't understand it, Andrew." So this time Wiles sent a fax trying to make the logical connection. Again Katz was not satisfied. Something was simply not right. This was supposed to be exactly one of the arguments that Wiles and Katz went over carefully in the spring when Wiles was teaching his "course." Any difficulties should have already been ironed out. But apparently the hole in Wiles' logic eluded them both. Possibly if the graduate students had stayed on, one of them might have alerted the two to the problem.

By the time Katz found the error, other mathematicians throughout the world were aware of the exact same problem with Wiles' proof. There simply was no Euler System here, and there was nothing doing. And without the Euler System—supposedly a generalization of the earlier work of Flach and Kolyvagin—there was no Class Number Formula. Without the Class Number Formula it was impossible to "count" the Galois representations of the elliptic curves against the modular forms, and Shimura-Taniyama was not established. And without the Shimura-Taniyama conjecture proved as correct, there was no proof of Fermat's Last Theorem. In short, the hole in the Euler System made everything collapse like a house of cards.

The Agony

Andrew Wiles returned to Princeton in the fall of 1993. He was embarrassed, he was upset, he was angry, frustrated, humiliated. He had promised the world a proof of Fermat's Last Theorem—but he couldn't deliver. In mathematics, as in almost any other field, there are no real "second prizes" or "also ran" awards. The crestfallen Wiles was back in his attic trying to fix the proof. "At this point, he was hiding a secret from the world," recalled Nick Katz, "and I think he must have felt pretty uncomfortable about it." Other colleagues had tried to help Wiles, including his former student Richard Taylor who was teaching at Cambridge but joined Wiles at Princeton to help him try to fix the proof.

"The first seven years, working all alone, I enjoyed every minute of it," Wiles recalled, "no matter how hard or seemingly impossible a hurdle I faced. But now, doing mathematics in this over-exposed way was certainly not my style. I have no wish to ever repeat this experience." And the bad experience lasted and lasted. Richard Taylor, his sabbatical leave over, returned to Cambridge and still Wiles saw no end in sight. His colleagues looked at him with a mixture of anticipation, hope, and pity, and his suffering was clear to everyone around him. People wanted to know. They wanted to hear good news, but none of his colleagues dared ask him how he was doing with the proof. Outside his department, the rest of the world was curious, too. Sometime in the night of December 4, 1993, Andrew Wiles posted an e-mail message to the computer news group Sci.math, to which several number theorists and other mathematicians belonged:

> In view of the speculation on the status of my work on the Taniyama-Shimura conjecture and Fermat's Last Theorem I will give a brief account of the situation. During the review process a number of problems emerged, most of which have been resolved, but one in particular I have not yet settled....I believe that I will be able to finish this in the near future using the ideas explained in my Cambridge lectures. The fact that a lot of work remains to be done on the manuscript makes it unsuitable for release as a preprint. In my course in Princeton beginning in February I will give a full account of this work.
>
> Andrew Wiles

The Post-Mortem

But Andrew Wiles was prematurely optimistic. And whatever course he may have planned to offer at Princeton would not yield any solution. When more than a year passed since his short-lived triumph in Cambridge, Andrew Wiles was about to give up all hope and to forget his crippled proof.

On Monday morning, September 19, 1994, Wiles was sitting at his desk at Princeton University, piles of paper strewn all around him. He decided he would take one last look at his proof before chucking it all and abandoning all hope to prove Fermat's Last Theorem. He wanted to see exactly what it was that was preventing him from constructing the Euler System. He wanted to know—just for his own satisfaction—why he had failed. Why was there no Euler System?— he wanted to be able to pinpoint precisely which technical fact was making the whole thing fail. If he was going to give up, he felt, then at least he was owed an answer to why he had been wrong.

Wiles studied the papers in front of him, concentrating very hard for about twenty minutes. And then he saw exactly why he was unable to make the system work. Finally, he understood what was wrong. "It was the most important moment in my entire working life," he later described the feeling. "Suddenly, totally unexpectedly, I had this incredible revelation. Nothing I'll ever do again will..." at that moment tears welled up and Wiles was choking with emotion. What Wiles realized at that fateful moment was "so indescribably beautiful, it was so simple and so elegant...and I just stared in disbelief." Wiles realized that exactly what was making the Euler System fail is what would make the Horizontal Iwasawa Theory approach he had abandoned three years earlier *work*. Wiles stared at his paper for a long time. He must be dreaming, he thought, this was just too good to be true. But later he said it was simply too good to be *false*. The discovery was so powerful, so beautiful, that it *had* to be true.

Wiles walked around the department for several hours. He didn't know whether he was awake or dreaming. Every once in a while, he would return to his desk to see if his fantastic finding was still there—and it was. He went home. He had to sleep on it—maybe in the morning he would find some flaw in this new argument. A year of pressure from the entire world, a year of one frustrated attempt after another had shaken Wiles' confidence. He came back to his desk in the morning, and the incredible gem he had found the day before was still there, waiting for him.

Wiles wrote up his proof using the corrected Horizontal Iwasawa Theory approach. Finally, everything fell perfectly into place. The approach he had used three years earlier was the correct one. And that knowledge came to him from the failing of the Flach and Kolyvagin route he had chosen in midstream. The manuscript was ready to be shipped out. Elated, Andrew Wiles logged into his computer account. He sent e-mail messages across the Internet to a score of mathematicians around the world: "Expect a Federal Express package in the next few days," the messages read.

As he had promised his friend Richard Taylor, who had come from England especially to help him correct his proof, the new paper correcting the Iwasawa theory bore both of their names even though Wiles obtained the actual result after Taylor's departure. Within the next few weeks, the mathematicians who received Wiles' correction to his Cambridge papers went over all the details. They could find nothing wrong.

Wiles now used the conventional approach to the presentation of mathematical results. Instead of doing what he had done in Cambridge a year and a half earlier, he sent the papers to a professional journal, the *Annals of Mathematics*, where they could be peer-reviewed by other mathematicians. The review process took a few months, but no flaws were found this time. The May, 1995, issue of the journal contained Wiles' original Cambridge paper and the correction by Taylor and Wiles. Fermat's Last Theorem was finally laid to rest.

Did Fermat Possess the Proof?

Andrew Wiles described his proof as "a twentieth century proof." Indeed, Wiles used the work of many twentieth-century mathematicians. He also used the work of earlier mathematicians. All the myriad elements of Wiles' constructions came from the work of others, many others. So the proof of Fermat's Last Theorem was really the achievement of a large number of mathematicians living in the twentieth century—and all the preceding ones to the time of Fermat himself. According to Wiles, Fermat could not possibly have had this proof in mind when his wrote his famous note in the margin. This much is true, of course, because the Shimura-Taniyama conjecture did not exist until the twentieth century. But could Fermat have had another proof in mind?

The answer is probably not. But this is not a certainty. We will never know. On the other hand, Fermat lived another 28 years after he wrote his theorem in the margin. And he never said anything more about it. Possibly he knew he couldn't prove the theorem. Or he may have erroneously thought that his method of infinite descent used in proving the simple case $n=3$ could apply to a general solution. Or maybe he simply forgot about the theorem and went on to do other things.

Proving the theorem the way it was finally done in the 1990s required a lot more mathematics than Fermat himself could have known. The profound nature of the theorem is that not only does its history span the length of human civilization, but the final solution of the problem came about by harnessing—and in a sense unifying—the entire breadth of mathematics. It was this unification of seemingly disparate areas of mathematics that finally nailed the theorem. And despite the fact that Andrew Wiles was the person who did the important final work on the theorem by proving a form of the Shimura-Taniyama conjecture needed to prove Fermat's theorem, the entire enterprise was the work of many people. And it is all their contributions, taken together, which brought about the final solution. Without the work of Ernst Kummer there would have been no theory of ideals, and without ideals the work of Barry Mazur would not have existed. Without Mazur there would have been no conjecture by Frey, and without the crucial conjecture and its synthesis by Serre there would not have been the proof by Ribet that the Shimura-Taniyama conjecture would establish Fermat's Last Theorem. And it seems that no proof of Fermat's Last Theorem would be possible without that conjecture put forward by Yutaka Taniyama at Tokyo-Nikko in 1955 and then refined and made specific by Goro Shimura. Or would it?

Fermat, of course, could not have formulated such an overarching conjecture that would unify two very different branches of mathematics. Or could he have done so? Nothing is certain. We only know that the theorem has finally been established and that the proof has been checked and verified to its finest details by scores of mathematicians throughout the world. But just because a proof exists and it is a very complicated, advanced one does not mean that a simpler proof is not possible. Ribet, in fact, points out in one of his papers a direction where a proof of Fermat's theorem might be possible without a proof of the Shimura-Taniyama conjecture. And perhaps Fermat did know a lot of powerful "modern" mathematics, now lost (actually, the copy of Bachet's Diophantus in which he supposedly wrote his margin-statement has never been found). So whether or not Fermat did possess a "truly marvelous proof" of his theorem, one that could not fit in the margin of his book, will forever remain his secret.

KENNETH A. RIBET

Left to right: John Coates, Andrew Wiles, Ken Ribet, Karl Rubin, celebrating Wiles' proof in Cambridge right after the historical presentation.

ROBERT P. MATTHEWS, PRINCETON UNIVERSITY

Gerd Faltings. He had a totally different approach to Fermat's Last Theorem. When Wiles failed in his first attempt in 1993, many feared that Faltings would now beat him to the true proof.

KENNETH A. RIBET

Andrew Wiles at the crucial moment of his third lecture at Cambridge, June 1993, when it was clear to everyone Fermat was just around the corner.

KENNETH A. RIBET

Ken Ribet at the famous café where he finished the proof that Shimura-Taniyama would imply that Fermat's Last Theorem had to be true.

Endnotes

1. E.T. Bell, *Men of Mathematics*, New York: Simon and Schuster, 1937, p. 56.

2. Barry Mazur, "Number Theory as Gadfly," *American Mathematical Monthly*, Vol. 98, 1991, p. 593.

3. Plimpton 322 and its implications about the advanced level of Babylonian mathematics were brought to the attention of the scientific community by Otto Neugebauer in 1934. An account in English can be found in his book *The Exact Sciences in Antiquity* (Princeton University Press, 1957).

4. Actually, Cantor went much farther. He hypothesized that the *order* of infinity of the irrational numbers immediately follows that of the rationals. That is, he believed that there is no order of infinity that is both higher than that of the rational numbers and lower than that of the irrational numbers. This statement became known as the Continuum Hypothesis, and the work of Kurt Gödel and Paul Cohen in the twentieth century established that it is impossible to prove this hypothesis within the rest of mathematics. The Continuum Hypothesis stands alone (with some equivalent restatements) opposite the rest of mathematics, their respective truths independent of each other. This remains one of the most bizarre truths in the foundations of mathematics.

5. D. Wells, *Curious and Interesting Numbers*, London: Penguin Books, 1987, p. 81.

6. C. Boyer, *A History of Mathematics*, New York: Wiley, 1968, p. 9.

7. Reprinted in B. Mazur, op. cit.

8. Ian Stewart, *Nature's Numbers*, New York: Basic Books, 1995, p. 140.

9. Michael Mahoney, *The Mathematical Career of Pierre de Fermat*, 2d ed., Princeton University Press, 1994, p. 4.

10. Harold M. Edwards, *Fermat's Last Theorem*, New York: Springer-Verlag, 1977, pp. 61-73.

11. Much of what is publicly known about the secret society comes from Paul R. Halmos, "Nicolas Bourbaki," *Scientific American*, 196, May 1957, pp. 88-97.

12. André Weil, *Oeuvres*, Vols. I-III, Paris: Springer-Verlag, 1979.

13. Adapted from Kenneth A. Ribet and Brian Hayes, "Fermat's Last Theorem and Modern Arithmetic," *American Scientist*, Vol. 82, March-April 1994, pp. 144-156.

14. A good introduction to this topic is the book by Joseph H. Silverman and John Tate, *Rational Points on Elliptic Curves*, New York: Springer-Verlag, 1992.

15. Most of the information on Yutaka Taniyama's life is from Goro Shimura, "Yutaka Taniyama and His Time: Very Personal Recollections," *Bulletin of the London Mathematical Society*, Vol. 21, 1989, pp. 184-196.

16. Reprinted in the Japanese journal *Sugaku*, May 1956, pp. 227-231.

17. Shimura thus stated to Serre his actual conjecture, sharing it for the first time, and implicitly trusting that Serre would acknowledge him as its originator.

18. André Weil, *Oeuvres*, op. cit., Vol. III, p. 450.

19. André Weil, "Über die Bestimmung Dirichletscher Reihen durch Funktionalgleichungen," *Math. Annalen*, Vol. 168, 1967, pp. 165-172.

20. Weil's letter to Lang, along with much of the chronology of events described here, including private conversations and letters, are reproduced in Serge Lang, "Some History of the Shimura-Taniyama Conjecture," *Notices of the American Mathematical Society*, November 1995, pp. 1301-1307. It is to Lang's credit that his article and the "Taniyama-Shimura File" he has been circulating among mathematicians for ten years now are finally helping bring Goro Shimura the recognition he rightly deserves.

21. Jean-Pierre Serre, "Lettre à J.-F. Mestre," reprinted in *Current Trends in Arithmetical Algebraic Geometry*, Providence: American Mathematical Society, 1987, pp. 263-268.

22. Barry Mazur, "Modular Curves and the Eisenstein Ideal," Paris, France: *The Mathematical Publications of I.H.E.S.*, Vol. 47, 1977, pp. 33-186.

23. Barry Mazur, op. cit.

24. The first and more important of the two papers, Andrew Wiles, "Modular Elliptic Curves and Fermat's Last Theorem," *Annals of Mathematics*, Vol. 142, 1995, pp. 443-551, begins with Fermat's actual margin-statement of his theorem in Latin: *Cubum autem in duos cubos, aut quadratoquadratum in duos quadratoquadratos, et generaliter nullam in infinitum ultra quadratum potestatem in duos ejusdem nominis fas est dividere: cujus rei demonstrationem mirabilem sane detexi. Hanc marginis exiguitas non caperet. Pierre de Fermat.* The journal sold out even before the publication date, and for the first time imposed a charge of $14 per individual issue.

Author's Note

In preparing this book I drew much of the historical background from a variety of sources. My favorite, and the most complete and original source, is E.T. Bell's book, *Men of Mathematics* (although I dislike the sexist title, which is also misleading since two of the mathematicians are women; the book was written in 1937). Apparently other historians of mathematics have drawn their information from Bell, so I will not mention them by name here. All my important sources are referenced in the endnotes. Additionally, I found the articles of Jacquelyn Savani of Princeton University (*Princeton Weekly Bulletin,* September 6, 1993) useful, and I thank her for sending me a copy of a program aired on the BBC about Fermat's Last Theorem.

I am indebted to C. J. Mozzochi for a number of photographs of mathematicians involved in the proof of Fermat's Last Theorem. Very warm thanks to Professor Kenneth A. Ribet of the University of California at Berkeley for informative interviews and much important information about his work leading to the proof of Fermat's theorem. My deep gratitude to Professor Goro Shimura of Princeton University for his generosity with his time in giving me access to so much important information about his work, and his conjecture without which there would be no proof of Fermat's theorem. I am grateful to Professor Gerd Faltings of the Max Planck Institute in Bonn and to Professor Gerhard Frey of the University of Essen, Germany, for provocative interviews and insightful opinions. Thanks also to Professor Barry Mazur of

Harvard University for explaining to me important concepts in the geometry of number theory. Any errors that remain are certainly mine.

I thank my publisher, John Oakes, for his encouragement and support. Thanks also to JillEllyn Riley and Kathryn Belden of Four Walls Eight Windows. And finally, my deep gratitude to my wife, Debra.

Index